Graham Jones, Steve Field, Chris Hewlett and David Styles

Cambridge International
AS & A Level

Physics

Practical Teacher's Guide

CAMBRIDGE
UNIVERSITY PRESS

University Printing House, Cambridge CB2 8BS, United Kingdom

One Liberty Plaza, 20th Floor, New York, NY 10006, USA

477 Williamstown Road, Port Melbourne, VIC 3207, Australia

314–321, 3rd Floor, Plot 3, Splendor Forum, Jasola District Centre, New Delhi - 110025, India

79 Anson Road, #06–04/06, Singapore 079906

Cambridge University Press is part of the University of Cambridge.

It furthers the University's mission by disseminating knowledge in the pursuit of education, learning and research at the highest international levels of excellence.

Information on this title: www.cambridge.org/ 9781108524902

First published 2018

20 19 18 17 16 15 14 13 12 11 10 9 8 7 6 5 4 3 2

Printed in Great Britain by CPI Group (UK) Ltd, Croydon CR0 4YY

A catalogue record for this publication is available from the British Library

ISBN 978-1-108-52490-2 Paperback

Acknowledgments

Cover image: David Parker/Science Photo Library

Contents

Introduction

Practical work is an essential part of any advanced physics course. For the Cambridge International AS & A Level Physics examinations, the assessment of practical skills in Paper 3 makes up 23% of the total marks at AS Level. Paper 3 and Paper 5 together make up 23% of the total marks at A Level.

The practical investigations in the Cambridge International AS & A Level Physics Practical workbook have been carefully chosen to:

- meet the requirements of all the learning objectives for specific practical activities
- provide progressive guidance and practice of Assessment Objective 3 (AO3) skills.

The order of the investigations presented follows the order of the topics in the Cambridge International AS & A Level Physics Coursebook, but please note that this does not mean that they must be completed in that order. Many of the investigations can be answered without knowing the particular theory, but it is hoped that you will find some investigations to enhance your teaching of the theory, while building up the confidence and ability of learners during the course.

The navigation grids summarise the practical skills that are assessed in Paper 3 (AS Level) and Paper 5 (A Level). You can use these grids to search for practical investigations that involve a particular skill or skills. At the beginning of each practical investigation, the learning objectives and skills that are supported are also listed.

These points have been provided to give extra support to students who may be struggling with the investigation.

These points provide additional tasks to extend more able learners.

Each chapter of the workbook offers more than one investigation so that you can choose those that suit the equipment and time that you have available. The apparatus required has been chosen to be as commonly available as possible and is largely listed as 'apparatus that is used regularly' in the syllabus.

Paper 3 requires candidates to undertake a practical examination and this may appear daunting at the start of the course. It is only by actually undertaking practical work and considering the problems, difficulties and safety aspects that learners will become confident and able to give their best effort in their examinations. Ideally, learners should work on their own, as they will do so in the examination, but during the course they may work in groups of two to provide mutual support and encouragement. The ultimate aim should be for learners to take readings and undertake the analysis for themselves. Paper 5 does not require an actual experiment to be undertaken. Most of the investigations where data is analysed in these later chapters have data that is given to learners. However, some investigations can be performed by the learners themselves and, where possible, you are encouraged to allow this to happen.

Although practical work requires time, it is time well spent. Practical work enables learners to acquire transferrable skills and gives them the confidence that the theory they have learned works in practice. Because of this, the details of the theory are more easily understood and retained. The important learning experiences while carrying out practical work are the range of skills that are being used and developed – the processes of planning, carrying out, observing, recording and analysing. The workbook gives the learners experience in developing these skills. It is **not** designed to be a series of mock practical examination papers! In carrying out the investigations, the learners will practise and acquire the skills that will enable them to be more confident when tackling the practical examination.

Learners should also be aware that there is guidance on practical skills for Paper 3 and Paper 5 in Chapters P1 and P2 of the coursebook in this series. You may like to use these chapters as an introduction or reference for learners.

Safety

Working safely in a physics laboratory is an essential aspect of learning which characterises practical work. It is the duty of the school to make it clear to learners just what is expected of them when they are working in a laboratory.

Many safety issues in a physics laboratory concern the prevention of damage to the equipment rather than to the learner.

Working with water	Place all the apparatus in a tray so that any spillage does not affect paperwork. If working with hot or boiling water, use tongs to handle containers such as beakers.
Using a liquid-in-glass thermometer	Place the thermometer securely on the bench, when not in use, so that it does not roll off the bench. If a thermometer breaks, inform the teacher immediately. Do not touch either the broken glass or the liquid from inside the thermometer.
Loading thin materials such as wires	Wear safety goggles in case of fracture of the wire. Beware of falling weights when the wire fractures and place a cushion or similar object on the floor.
Connecting electrical components	Do not exceed the recommended voltage for the component: for example, a 6 V lamp.
Toppling retort stands	If a stand is moving or in danger of toppling, secure it to the bench using a G-clamp.
Rolling objects such as cylinders	Place a suitable object such as a box to collect the object so that it does not fall to the floor or affect somebody else's experiment.
Dry cells such as 1.5 V batteries	Do not connect the terminals of the cell to each other with a wire.
Using sharp blades or pins	Tape over sharp edges; keep points of pins downwards, away from eyes.

Table S1

AS Level Practical Skills

The following grids map the practical investigations from the workbook to the mark categories for Papers 3 and 5, as listed in the Cambridge International AS & A Level Chemistry syllabus.

The grids are designed to aid you when planning practical and theory lessons, to ensure learners develop the practical skills required as part of this course.

Manipulation, measurement and observation (MMO)

SKILL	CHAPTER							
	1	2	3	4	5	6	7	8
Successful collection of data								
(a) Set up apparatus, follow instructions and make measurements using common laboratory apparatus	1.1; 1.2; 1.3; 1.4	2.1; 2.2; 2.3	3.1; 3.2; 3.3; 3.4	4.1; 4.2; 4.3; 4.4	5.1; 5.2; 5.3	6.1; 6.2; 6.3	7.1; 7.2; 7.3	8.1; 8.2; 8.3
(b) Repeat readings where appropriate	1.2; 1.4		3.1; 3.2; 3.3; 3.4		5.2; 5.3	6.1; 6.2; 6.3	7.1; 7.2; 7.3	
(c) Set up a circuit from a circuit diagram	1.3					6.1; 6.2; 6.3	7.1; 7.2; 7.3	
(d) Use a multimeter	1.3							
(e) Use a micrometer	1.3				5.2		7.3	
(f) Measure time intervals	1.4	2.1	3.1; 3.4		5.2; 5.3			
(g) Measure the period of an oscillating system	1.1; 1.2; 1.3; 1.4	2.1; 2.2; 2.3	3.1; 3.2; 3.3; 3.4	4.1; 4.3; 4.4	5.1; 5.3	6.1; 6.2; 6.3	7.1; 7.2; 7.3	8.1; 8.2; 8.3
(h) Collect an appropriate number of sets of data, make measurements that span the largest possible range	1.1; 1.2; 1.3; 1.4	2.1; 2.2; 2.3	3.1; 3.2; 3.3; 3.4	4.1; 4.3; 4.4	5.1; 5.2; 5.3	6.1; 6.2; 6.3	7.1; 7.2; 7.3	8.1; 8.2; 8.3

Presentation of data and observations (PDO)

SKILL	CHAPTER							
	1	2	3	4	5	6	7	8
Presentation of data								
(a) Present numerical data in a single table of results, Include columns for raw data and for values calculated from them	1.1; 1.2; 1.3; 1.4	2.1; 2.2; 2.3	3.1; 3.2; 3.3; 3.4	4.1; 4.3; 4.4	5.1; 5.2; 5.3	6.1; 6.2; 6.3	7.1; 7.2; 7.3	8.1; 8.2; 8.3
(b) Use column headings that include both quantity and unit	1.1; 1.2; 1.3; 1.4	2.1; 2.2; 2.3	3.1; 3.2; 3.3; 3.4	4.1; 4.3; 4.4	5.1; 5.2; 5.3	6.1; 6.2; 6.3	7.1; 7.2; 7.3	8.1; 8.2; 8.3
(c) Record raw readings of a quantity to the same precision	1.1; 1.2; 1.3; 1.4	2.1; 2.2; 2.3	3.1; 3.2; 3.3; 3.4	4.1; 4.3; 4.4	5.1; 5.2; 5.3	6.1; 6.2; 6.3	7.1; 7.2; 7.3	8.1; 8.2; 8.3
(d) Use and justify the correct number of significant figures in calculated quantities	1.1; 1.2; 1.3; 1.4	2.2	3.1; 3.2; 3.3; 3.4	4.1; 4.2; 4.3	5.1; 5.2; 5.3	6.1; 6.2; 6.3	7.1; 7.2; 7.3	8.1; 8.2; 8.3

SKILL	CHAPTER							
	1	2	3	4	5	6	7	8
Graphs								
(a) Label axes with both quantity and unit	1.1; 1.2; 1.3; 1.4	2.2; 2.3	3.1; 3.2; 3.3	4.1; 4.3	5.1; 5.3			8.1; 8.2; 8.3
(b) Choose scales for axes such that the data points occupy at least half of the graph grid in both *x*- and *y*-directions	1.1; 1.2; 1.3; 1.4	2.2; 2.3	3.1; 3.2; 3.3	4.1; 4.3	5.1; 5.3		7.3	8.1; 8.2; 8.3
(c) Use a false origin where appropriate	1.1; 1.2; 1.3; 1.4	2.2; 2.3	3.1; 3.2; 3.3	4.1; 4.3	5.1; 5.3		7.3	8.1; 8.2; 8.3
(d) Choose scales for axes that allow the graph to be read easily	1.1; 1.2; 1.3; 1.4	2.2; 2.3	3.1; 3.2; 3.3	4.1; 4.3	5.1; 5.3		7.3	8.1; 8.2; 8.3
(e) Place regularly spaced numerical labels along each axis	1.1; 1.2; 1.3; 1.4	2.2; 2.3	3.1; 3.2; 3.3	4.1; 4.3	5.1; 5.3		7.3	8.1; 8.2; 8.3
(f) Plot data points to an accuracy of better than 1 mm	1.1; 1.2; 1.3; 1.4	2.2; 2.3	3.1; 3.2; 3.3	4.1; 4.3	5.1; 5.3	6.1; 6.2; 6.3	7.1; 7.2; 7.3	8.1; 8.2
(g) Draw straight best-fit lines or curves to show the trend of a graph	1.1; 1.2; 1.3; 1.4	2.2; 2.3	3.1; 3.2; 3.3	4.1; 4.3	5.1; 5.3	6.1; 6.2; 6.3	7.1; 7.2; 7.3	8.1; 8.2
(h) Draw tangents to curved trend lines					5.3			

Analysis, conclusions and evaluation (ACE)

SKILL	CHAPTER							
	1	2	3	4	5	6	7	8
Interpretation of graphs								
(a) Relate straight-line graphs to equations such as $y = mx + c$ and derive expressions that equate to the gradient and intercept	1.1; 1.2; 1.4		3.2; 3.3	4.1; 4.3		6.2; 6.3	7.1; 7.2; 7.3	8.1; 8.2
(b) Read the coordinates of points on the trend line	1.1; 1.2; 1.3; 1.4	2.3	3.1; 3.2; 3.3	4.1; 4.3	5.1	6.2; 6.3	7.1; 7.2; 7.3	8.1; 8.2
(c) Determine the gradient of a straight-line graph or a tangent	1.1; 1.2; 1.3; 1.4		3.1; 3.2; 3.3	4.1; 4.3	5.1; 5.3	6.2; 6.3	7.1; 7.2; 7.3	8.1; 8.2
(d) Determine the *y*-intercept of a straight-line graph			3.1; 3.2; 3.3	4.1; 4.3	5.1	6.2; 6.3	7.1; 7.2; 7.3	8.1; 8.2

SKILL	CHAPTER							
	1	2	3	4	5	6	7	8
Drawing conclusions								
(a) Draw conclusions from an experiment including determining the values of constants, considering whether experimental data supports a given hypothesis, and making predictions	1.1; 1.2; 1.3; 1.4	2.1; 2.2	3.1; 3.2; 3.3; 3.4	4.1; 4.3	5.2; 5.3	6.1; 6.2; 6.3	7.1; 7.2; 7.3	8.1; 8.2; 8.3
Estimating uncertainties								
(a) Estimate, quantitatively, the uncertainty in a measurement	1.2		3.4		5.2	6.3		8.3
(b) Determine the uncertainty in a final result								
(c) Express the uncertainty as an absolute, fractional or percentage uncertainty	1.4		3.4			6.3		8.3
(d) Express the uncertainty in a repeated measurement as half the range of the readings	1.2		3.4					
Identifying limitations								
(a) Identify and describe limitations in an experimental procedure	1.1; 1.2	2.1; 2.2; 2.3		4.2	5.3	6.2	7.2	8.3
(b) Identify the most significant sources of uncertainty in an experiment				4.2; 4.4	5.3		7.2; 7.3	8.3
(c) Show an understanding of the distinction between systematic and random errors							7.1	
Suggesting improvements								
(a) Suggest modifications to an experimental arrangement that will improve the accuracy. Describe these modifications clearly in words or diagrams	1.1	2.1; 2.2; 2.3		4.4	5.1; 5.3			8.3

A Level Practical Skills

SKILL	CHAPTER							
	9	10	11	12	13	14	15	16
Planning								
(a) Defining the problem, variables and controls	9.1, 9.3, 9.5, 9.7	10.2	11.2, 11.4v	12.3	13.1, 13.3, 13.5	14.2, 14.3	15.1, 15.3, 15.4	16.3
(b) Describe methods to measure, change and control variables	9.3, 9.5, 9.6, 9.7	10.2	11.2	12.3	13.1, 13.3, 13.5	14.2, 14.3	15.1, 15.3, 15.4	16.3
(c) Method of analysis; explain how a proposed relationship is analysed	9.1, 9.3, 9.5, 9.7	10.2	11.2, 11.4	12.3	13.1, 13.3, 13.5	14.2, 14.3	15.1, 15.3, 15.4	16.3
(d) Give detail in planning an experiment, including safety, assessing risks, describing precautions and giving detailed use of apparatus	9.3, 9.5, 9.7	10.2	11.2	12.3	13.1, 13.3, 13.5	14.2, 14.3	15.1, 15.3, 15.4	16.3
Analysis, conclusion and evaluation								
(a) Draw an appropriate results table	9.3, 9.4, 9.5, 9.6, 9.7	10.2	11.2	12.3	13.1, 13.2, 13.3, 13.4, 13.5, 13.6	14.1, 14.2, 14.3, 14.4, 14.5	15.1, 15.2, 15.3, 15.4, 15.5	16.3
(b) Use logarithms or exponentials	9.6	10.4	11.1, 11.2, 11.3, 11.4		13.2, 13.3			16.2, 16.3
(c) Draw a graph, including matching axes to an equation	9.4, 9.6	10.1, 10.3, 10.4, 10.5	11.1, 11.3, 11.4	12.1, 12.2, 12.3, 12.4	13.2, 13.4, 13.6	14.1, 14.4, 14.5	15.2, 15.5	16.1, 16.2, 16.3, 16.4
(d) Draw error bars and a worst acceptable line	9.4, 9.6	10.1, 10.3, 10.4, 10.5	11.1, 11.3, 11.4	12.1, 12.2, 12.4	13.2, 13.4, 13.6	14.1, 14.4, 14.5	15.2, 15.5	16.1, 16.2, 16.4
(e) Analyse a graph including gradient and intercept	9.1, 9.3, 9.4, 9.5, 9.6, 9.7	10.1, 10.3, 10.4, 10.5	11.1, 11.3, 11.4	12.1, 12.2, 12.4	13.1, 13.2, 13.3, 13.4, 13.5, 13.6	14.1, 14.2, 14.3, 14.4, 14.5	15.1, 15.2, 15.3, 15.4 15.5	16.1, 16.2, 16.3, 16.4
(f) Use, combine and calculate uncertainties	9.2, 9.4, 9.6	10.1,10.3, 10.4	11.1, 11.3, 11.4	12.1, 12.2, 12.4	13.2, 13.4, 13.6	14.1, 14.4, 14.5	15.2, 15.5	16.1, 16.2, 16.4
g) Evaluation of experimental techniques and the effect of uncertainties	9.4, 9.6	10.1, 10.2, 10.3, 10.4	11.1, 11.2, 11.3, 11.4	12.1, 12.2, 12.3, 12.4	13.2, 13.4, 13.6	14.1, 14.4, 14.5	15.2, 15.5	16.1, 16.2, 16.3, 16.4

Chapter 1:
Using apparatus

Chapter outline

This chapter relates to Chapter 1: Kinematics – describing motion, Chapter 7: Matter and materials and Chapter 9: Electric current, potential difference and resistance, in the coursebook.

In this chapter learners will complete investigations on:

- 1.1 Determining the density of water
- 1.2 Determining the spring constant of a spring
- 1.3 Determining the resistance of a metal wire
- 1.4 Determining the average speed of a cylinder rolling down a ramp.

Practical Investigation 1.1: Determining the density of water

Skills focus

See the Skills grids at the front of this book for details of the skills developed and used in this investigation.

Duration

The practical work will take 30 minutes; the analysis and evaluation questions will take 30 minutes.

Preparing for the investigation

- Learners should be able to recall and use the equation $\rho = \frac{m}{V}$.
- In Part 1, learners are asked to record the mass of water for a particular volume.
- In Part 2, learners will measure the diameter d of the beaker. They will change the height h of the water in the beaker and plot their results as a graph of mass against h. The density of water can be found from the gradient of the graph.
- This practical can be conducted at the start of a course and does not have to coincide with the teaching of the theory.

Equipment

Each learner or group will need:

- metre rule
- 30 cm ruler
- 250 cm^3 beaker
- Vernier or digital callipers.

Access to:

- jug of water
- top-pan balance.

Safety considerations

- Clear any spillages of water.

Carrying out the investigation

- Learners may find it difficult to measure the height of the water in the beaker because the 30 cm ruler does not start from zero and may not be held vertically. Using a clamped metre rule is a better option.
- Learners should be aware that reading from a clamped metre rule could involve a parallax error because the rule cannot be placed close to the water surface.
- Learners may find it difficult to measure the height of the water for a number of other reasons including the shape of the meniscus and the curvature and thickness of the bottom of the beaker.
- Learners may find it difficult to measure the diameter of the beaker because it may vary, for example, where the lip of the beaker opens out. Also, placing the ruler or rule exactly across a diameter is difficult.

Some learners may need help choosing a scale to make good use of the graph grid. To help them gain confidence with their graph skills, produce some data that will prove challenging to plot on a 24 cm × 16 cm grid. An example is given in Table 1.1.

h / cm	m / g
2.1	72.1
2.9	95.8
3.8	121.8
4.6	144.1
5.6	173.9
6.4	196.2

Table 1.1

- If a learner needs to redraw their graph, supply a graph grid of identical size (24 cm × 16 cm) to the one in the workbook. Learners can insert the new grid in the appropriate place in the workbook.

Learners should consider how close their values are to the accepted value for the density of water. They should research factors that may affect the value of the density of water (e.g. temperature), and consider whether these factors account for any discrepancy.

Sample results

Table 1.2 provides sample results learners may obtain in the investigation.

d = 6.6 cm (using metre rule) = 6.612 cm (using digital callipers) Mass of beaker = 99.0 g

h / cm	Balance reading / g	m / g
2.1	167.1	68.1
3.1	195.6	96.6
4.1	250.4	151.4
5.7	281.2	182.2
6.7	320.8	221.8
7.8	357.7	258.7

Table 1.2

Answers to the workbook questions (using the sample results)

a See Table 1.2.

b, c See Figure 1.1.

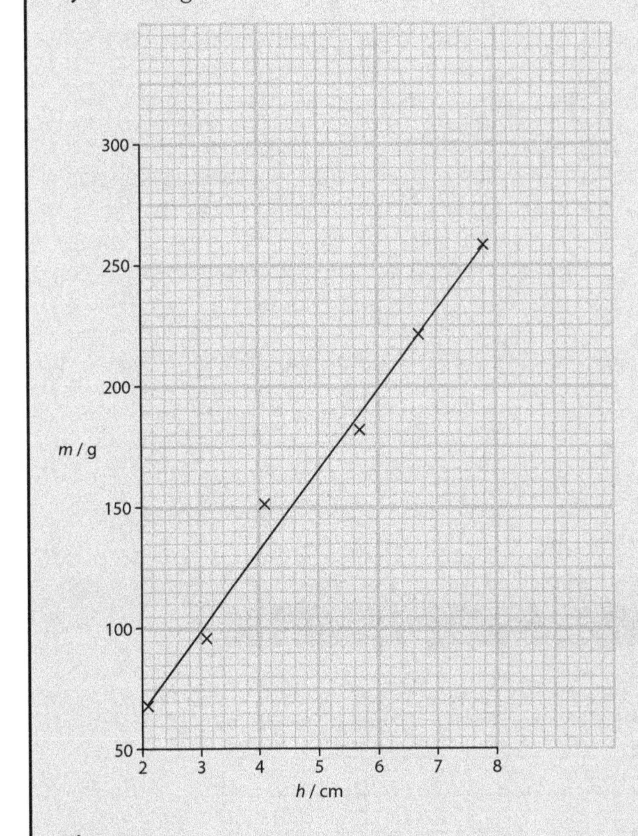

Figure 1.1

d Gradient = 33.04

e $m = \rho V$ so $m = \dfrac{\rho \pi d^2 h}{4}$

f Proof

g $d = 6.6$ cm so $\rho = \dfrac{4 \times 33.04}{\pi \times 6.6^2} = 0.966\ \text{g cm}^{-3}$

h Advantages of callipers:

- able to measure the inside diameter of the beaker
- more certain of measuring the maximum distance between opposite sides of the beaker.

Practical Investigation 1.2: Determining the spring constant of a spring

Skills focus

See the Skills grids at the front of this book for details of the skills developed and used in this investigation.

Duration

The practical work will take 30 minutes; the analysis and evaluation questions will take 30 minutes.

Preparing for the investigation

- Learners need to be able to recall and use the equation $k = \frac{F}{e}$.
- In Part 1, learners are asked to determine the extension of a spring for a particular mass suspended from the spring.
- In Part 2, learners determine the extension of the spring when a newton-meter is attached to the spring. They change the position of the newton-meter and the force F acting on the spring and plot their results as a graph of extension against F. The spring constant of the spring can be found from the gradient of the graph.
- This practical can be conducted at the start of a course and does not have to coincide with the teaching of the theory.

Equipment

Each learner or group will need:

- expendable steel spring
- 100 g mass hanger
- 0–10 N newton-meter
- 30 cm ruler
- four 100 g slotted masses
- two stands
- two bosses
- two clamps
- G-clamp.

Safety considerations

- Learners should take care when the bottom clamp is moved because the newton-meter and/or the spring could roll off the end of the rod.

Carrying out the investigation

- Learners may find it difficult to measure the length of the coiled section of spring because two positions of the rule must be viewed at the same time. The exact positions where the coiled section starts and ends are unclear and the spring may be moving while measurements are being taken.
- Learners should be made aware that any zero error in the newton-meter could result in a systematic error in the readings unless the zero error is taken into account.

Some learners may need help choosing a scale to make good use of the graph grid.

Data on springs would normally be given in N m^{-1}. Learners will probably have measured in centimetres and newtons. Encourage them to represent their results in other units, such as kN m^{-1} or N mm^{-1}.

Learners should discover the accepted value for k. They could look at the catalogue that the springs were purchased from.

Learners could repeat the experiment to see if all the springs from the same batch have the same value of k.

Learners could research which metals springs are made from.

When the topic is dealt with later on in the course, learners could debate whether this is a suitable method for studying the elastic properties of a material. Would it work for a thick metal rod?

Challenge more able learners to suggest why force–extension (and stress–strain) graphs are normally shown with extension as the independent variable plotted along the x-axis and force as the dependent variable along the y-axis? (**Answer**: In a tensile-testing machine, the sample is extended by a known amount and the resulting tension in the sample is measured.)

Common learner misconceptions

- Learners may incorrectly assume that the value of a slotted mass is equal to the value marked on it.

Sample results

Table 1.3 provides sample results.

$x_0 = 2.0\text{ cm}$

F / N	x / cm	e / cm
1.0	5.6	3.6
2.0	10.1	8.1
3.0	14.4	12.4
4.0	18.7	16.7
5.0	22.8	20.8
6.0	27.0	25.0

Table 1.3

Answers to the workbook questions (using the sample results)

a See Table 1.3.

b, c See Figure 1.2.

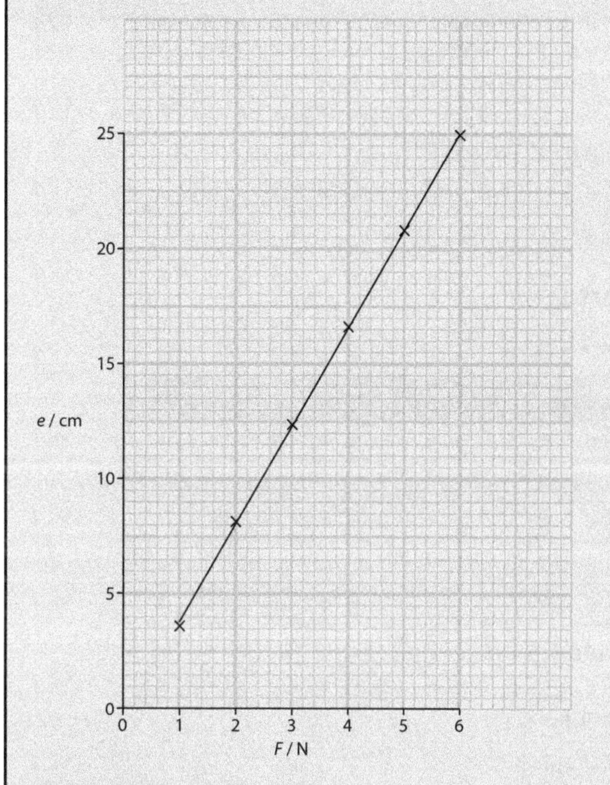

Figure 1.2

d 4.27 (cm N^{-1})

e Gradient $= \frac{e}{F}$ and $k = \frac{F}{e}$ so $k = \frac{1}{\text{gradient}}$

f $k = \frac{1}{\text{gradient}} = 23.4\,N\,m^{-1}$

g If x_0 is larger, values of x would be smaller, but it is the difference between the x values that determines the gradient. However, the value of x_0 has probably become progressively bigger throughout the experiment so this would lead to a smaller gradient and a larger value of k.

h Smaller gradient, same y-intercept (0, 0).

i See Table 1.4

Advantages	Disadvantages
measures in newtons	zero error
loaded spring is stable when measuring length	small scale so difficult to estimate fractions of a newton

Table 1.4

Practical Investigation 1.3: Determining the resistance of a metal wire

Skills focus

See the Skills grids at the front of this book for details of the skills developed and used in this investigation.

Duration

The practical work will take 30 minutes; the analysis and evaluation questions will take 30 minutes.

Preparing for the investigation

- Learners need to be able to recall and use the equation $R = \frac{V}{I}$.
- In Part 1, learners are asked to tape a length of constantan wire to a metre rule. They set up a circuit using a cell, switch, two digital multimeters, connecting leads, crocodile clips and the wire from a circuit diagram. They determine the resistance (resistance per unit length) of the wire using one reading from each meter.
- In Part 2, learners use the rheostat to vary the potential difference across the wire and plot their results as a graph of current I against potential difference V. The resistance of the wire can be found from the gradient of the graph. The learners measure the diameter of the wire and use the diameter and the resistance of the wire to identify it from a table of wires with different swg.
- The jaws of the crocodile clips should be cleaned so that they make a good electrical contact with the resistance wire.
- This practical can be conducted at the start of a course and does not have to coincide with the teaching of the theory.

Equipment

Each learner or group will need:

- 1.5 V cell
- connecting leads
- crocodile clips
- power supply
- two digital multimeters
- rheostat

- metre rule
- switch.

Access to:

- reel of 36 swg constantan wire
- scissors
- adhesive tape
- wire cutters
- micrometer.

Safety considerations

The switch in the circuit should be opened between readings so that the battery does not become discharged and the circuit components do not become overheated.

Carrying out the investigation

- Learners should be aware that meter readings fluctuate causing uncertainties in readings.
- The learner should be aware that moving the slider on the rheostat through regular distances does **not** produce a set of evenly spaced readings.
- The learner should be aware that the connecting wires contribute towards the resistance of the circuit (see Investigation 7.1).

  Some learners may need help choosing a scale to make good use of the graph grid.

  Learners will probably have measured in milliamps and volts. Encourage them to represent their results in amps to give the final value in ohms.

  If a learner has a negative reading on a meter, explain why this has happened. Identify the positive and negative terminals of the meters.

  Data on resistance would normally give the tolerance as a percentage and the maximum power that the wire can withstand. Learners could calculate if their result is within 5% of the value of the swg (once revealed) and also calculate the power using $V \times I$.

Common learner misconceptions

- Learners may assume the length of the wire is the same as the distance between the crocodile clips. This might lead to an error if the wire is not straight and therefore longer.

Sample results

Tables 1.5 and 1.6 provide sample results learners may obtain in the investigation.

Part 1

See Table 1.5

Scale	Reading
600 V	000
200 V	01.4
20 V	1.49
2000 mV	1510
200 mV	1

Table 1.5

Part 4

See Table 1.6

V / V	I / A
0.77	0.0442
0.87	0.0494
0.95	0.0538
1.04	0.0591
1.13	0.0640
1.19	0.0674

Table 1.6

Answers to the workbook questions (using the sample results)

Part 2

3 i Six connecting leads

ii Two crocodile clips

Part 3

See Table 1.7

Connections	Does the resistance reading change when the slider is moved?
A and B	yes
B and C	no
A and C	yes

Table 1.7

Part 4

a, b See Figure 1.3.

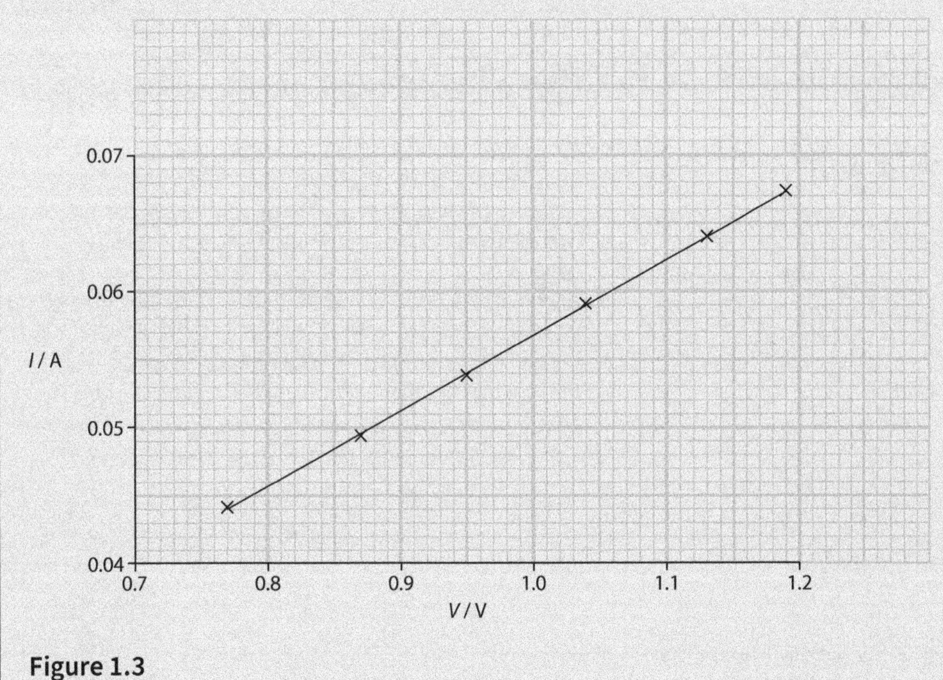

Figure 1.3

c 0.0556

d 18.0 Ω

e Measure the length of rod along which the slider moves and divide it into six equal lengths for the positions of the slider.

Part 5

a Diameter = 0.19 mm and resistance per unit length = 18 Ω; wire C

b R = 18 Ω and d = 0.19 mm suggest the wire is C (16.8 Ω, 0.19 mm)

c Smaller diameter → bigger resistance → smaller gradient.

Practical Investigation 1.4: Determining the average speed of a cylinder rolling down a ramp

Skills focus

See the Skills grids at the front of this book for details of the skills developed and used in this investigation.

Duration

The practical work will take 30 minutes; the analysis and evaluation questions will take 30 minutes.

Preparing for the investigation

- Learners should be able to recall and use the equation:

$$\text{average speed} = \frac{\text{distance travelled}}{\text{time taken}}$$

- In Part 1 learners are asked to investigate their reaction time and calculate the percentage uncertainty in a stopwatch reading.
- In Part 2 learners determine the average speed v of the cylinder as it rolls down the inclined plane using measurements of distance and time.
- In Part 3 learners are asked to measure the angle θ between the board and the bench.
  Learners vary θ and take readings of θ and time t and plot their results as a graph of v against $t\sin\theta$.
  The acceleration due to gravity g can be found from the gradient of the graph.
- Learners will appreciate in Part 1 what percentage of measured time is human reaction time.
- This practical can be conducted at the start of a course and does not have to coincide with the teaching the theory.

Equipment

Each learner or group will need:

- wooden cylinder of approximate diameter 2 cm and approximate length 10 cm
- wooden board of approximate dimensions 100 cm × 20 cm × 2 cm
- stand
- boss
- clamp
- metre rule
- protractor
- stopwatch
- book or pencil case to act as a barrier at the bottom of the ramp.

Safety considerations

- Learners should use the book or pencil case to stop the cylinder after it has reached the bottom of the wooden board.

Carrying out the investigation

- The cylinder does not always follow the same path down the slope.
- Friction between the cylinder and the board has an effect on the value of g. Will it be bigger or smaller than the accepted value?
- There is more scatter on this graph than on the other graphs drawn in this chapter.
- Learners should be aware that distance travelled must remain constant throughout.
- Learners should be aware that there is a maximum value of θ before t is too short to measure.

  Some learners may need help choosing a scale to make good use of the graph grid.
- Learners will probably have measured in centimetres. Encourage them to record their value(s) of distance in metres to an appropriate precision, e.g. 0.986 m.

  Some factors such as latitude and altitude affect the value of g. Learners could consider if any of these factors locally mean that the predicted value of g will be far from the accepted value of 9.81 m s^{-2}. Some learners may comment that they use the value of 9.8 in A Level Mechanics. This is for convenience because 9.8 is divisible by 7.

Sample results

Table 1.8 provides sample results learners may obtain in the Part 3 investigation.

θ / °	sin θ	t_1 / s	t_2 / s	t_3 / s	t_{mean} / s	$t\sin\theta$ / s	v / cm s^{-1}
4	0.070	2.34	2.34	2.31	2.33	0.16	42.9
5	0.087	2.16	2.13	2.22	2.17	0.19	46.1
8	0.14	1.56	1.57	1.62	1.58	0.22	63.3
10	0.174	1.53	1.47	1.56	1.52	0.264	65.8
12	0.208	1.34	1.37	1.40	1.37	0.285	73.0
14	0.242	1.28	1.25	1.25	1.26	0.305	79.4

Table 1.8

Answers to the workbook questions (using the sample results)

Part 1

a See Table 1.9

t_1 / s	t_2 / s	t_3 / s
0.19	0.16	0.22

Table 1.9

Mean value of $t = 0.19$ s

b 1.44 ± 0.03 so percentage uncertainty $= \frac{0.03}{1.44} \times 100 =$ 2.1%

Part 2

a 1.37 s, 1.47 s, 1.43 s so mean $t = 1.42$ s

b $d = 99.2$ cm so average speed $= \frac{99.2}{1.42} = 69.9$ cm s^{-1}

Part 3

b, c See Figure 1.4.

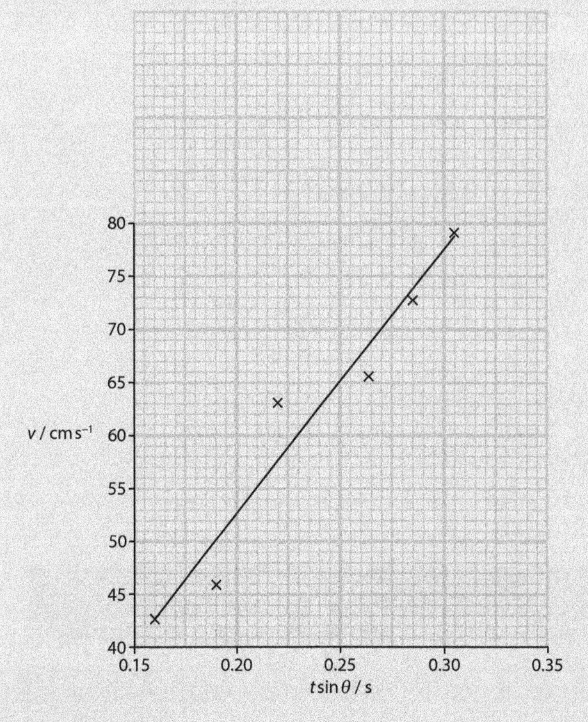

Figure 1.4

d Gradient = 252

e 7.56 m s^{-2}

f Yes by 225 cm s^{-2}

g No: the *y*-intercept is 1.88

h No, there is scatter about a best-fit straight line using all plotted points.

Chapter 2: Limitations and improvements

Chapter outline

This chapter relates to Chapter 4: Forces – vectors and moments and Chapter 7: Matter and materials, in the coursebook.

In this chapter learners will complete investigations on:

- 2.1 Thermal energy loss from water in a polystyrene cup
- 2.2 Loaded rubber band
- 2.3 Balanced metre rule.

Practical Investigation 2.1: Thermal energy loss from water in a polystyrene cup

Skills focus

See the Skills grids at the front of this book for details of the skills developed and used in this investigation.

Duration

The practical work will take about 30 minutes; the analysis and evaluation questions will take about 30 minutes.

Preparing for the investigation

- Learners will investigate how the rate of temperature decrease of hot water depends on the mass of the water.
- Learners will be taking measurements using a thermometer and a stopwatch.
- Learners will consider limitations of the procedure and suggest improvements.
- This practical can be conducted at the start of a course and does not have to coincide with the teaching of the theory.

Equipment

Each learner or group will need:

- long stem thermometer: −10 °C to 110 °C × 1 °C
- 200 cm^3 polystyrene cup
- stopwatch
- stirrer
- paper towel.

Access to:

- electric kettle or other means to heat water to boiling safely
- top-pan balance
- jug of cold water
- waterproof pen.

Safety considerations

- Students should take care when using hot water.
- When the thermometer is not in use it should be placed on a paper towel so that it does not fall onto the floor.

Carrying out the investigation

When discussing limitations and improvements explain why the improvements listed in Table 2.1 are not relevant or appropriate.

Limitation	Inappropriate improvement (and reason)
The lines on the cup are not equally spaced.	Provide a ruler. (It is not necessary for the lines to be evenly spaced to conduct this investigation.)
Heat was lost from the surface of the water.	Provide a lid for the cup. (The investigation is about heat loss from the surface of the liquid.)

Table 2.1

- Explain that some limitations cannot be addressed by an improvement but it is still justifiable to mention them giving a reason; for example, 'it was difficult to draw the bottom line as a straight line because I could not see what I was doing from above the cup'.

Sample results

The learners' results should be similar to the data below.

Mass of cup = 2.2 g

Mark on cup	Mass of cup and water / g	*m* / g	*t* / s for starting temperature 85 °C	80 °C
bottom	33.1	30.9	39.63	45.03
middle	75.4	73.2	59.94	75.84
top	118.8	116.6	82.28	101.00

Table 2.2

Answers to the workbook questions (using the sample results)

a See the values in the completed table of results, Table 2.2.

b *t* increases as *m* increases.

c The same starting temperature and temperature change were used because heat loss depends on the difference between the temperature of the object and room temperature.

d See Table 2.2. If the starting temperature is lower, the times will be greater because the excess temperature is lower and the rate of heat loss will be less. The times will be in the same order.

e There are many possible answers. Table 2.3 provides one possible limitation and improvement.

	Limitation	Improvement
E	The lines drawn on the cup are not straight / too thick.	Use a finer pen.

Table 2.3

Practical Investigation 2.2: Loaded rubber band

Skills focus

See the Skills grids at the front of this book for details of the skills developed and used in this investigation.

Duration

The practical work will take 30 minutes; the analysis and evaluation questions will take 30 minutes.

Preparing for the investigation

- In Part 1 learners will suspend a rubber band from two rods and attach a mass to the bottom of the rubber band. They will investigate how the extension of the rubber band varies with the separation of the rods.
- In Part 2 learners will suspend the same loaded rubber band from one rod and compare the extension of the rubber band with that in Part 1.
- Learners will consider limitations of the procedure and suggest improvements.
- This practical can be conducted at the start of a course and does not have to coincide with the teaching of the theory.

Equipment

Each learner or group will need:

- two stands
- two bosses
- two clamps
- G-clamp
- 100 g mass hanger
- four 100 g slotted masses
- protractor
- metre rule
- rubber band with approximate cross-section 2 mm × 1 mm and approximate circumference 20 cm.

Safety considerations

- Learners should take care when moving the stand. It may topple when the separation of the stands is large.
- Learners should not extend the rubber band too much. This could fracture the rubber band, causing the masses to fall to the bench or the floor.

Carrying out the investigation

When discussing limitations and improvements, explain that the improvement given in Table 2.4 is **not** appropriate because it could have been achieved with the existing apparatus.

Limitation	Improvement
The bottom of the mass rested on the bench.	Raise the height of the bosses and clamps.

Table 2.4

Sample results

The learners' results should be similar to the data below.

Part 1: Suspending the rubber band from two rods

Unextended length of rubber band $C = 20.0$ cm

x / cm	θ / °	$(\theta/2)$°	sin $(\theta/2)$	L / cm	e / cm
10.0	42	21	0.358	37.9	17.9
13.1	46	23.0	0.391	46.6	26.6
16.9	56	28.0	0.469	52.9	32.9
20.4	71	35.5	0.580	55.5	35.5

Table 2.5

Part 2: Suspending the rubber band from one rod

When the rubber band is suspended from one clamp, $R = 21.6$ cm

Answers to the workbook questions (using the sample results)

Part 1: Suspending the rubber band from two rods

a See the completed Table 2.5.

b L increases as x increases.

c See Table 2.5.

Part 2: Suspending the rubber band from one rod

a $e = 2R - C = (21.6 \times 2) - 20 = 23.2$ cm

b $x \approx 12$ cm

Learners could answer this at different levels.

Looking at the data in Table 2.5 suggests that because 23.2 cm is between 17.9 cm and 26.6 cm the value of x is between 10.0 cm and 13.1 cm.

If learners assume that there is a linear relationship, then the following calculation can be used:

$26.6 - 17.9 = 8.7$

$(23.2 - 17.9 = 5.3)$

Take $\frac{5.3}{8.7} \times (13.1 - 10)$ added to 10.0

This equals 11.88, which is approximately 12 cm.

Learners could plot a graph of e against x as shown in Figure 2.1, and extrapolate for an extension of 23.2 cm.

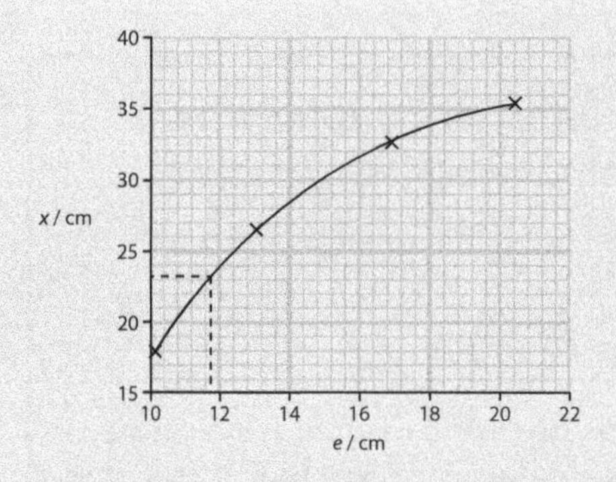

Figure 2.1

c Possible limitations and improvements are shown in Table 2.6.

Limitation	Improvement
The properties of the rubber band could change as a result of loading. It could be permanently deformed.	Measure C before the experiment.
The moveable stand toppled over for large values of x.	Use another G-clamp.
It was awkward to use a metre rule to measure x because the rule was too big	Use a 30 cm ruler instead.

Table 2.6

The suggestions should be realistic and achievable in a school laboratory. They could relate to either the apparatus, the experimental procedure or the sources of error learners have identified. If the learners needed to make improvements whilst carrying out the investigation, these could also be included.

Practical Investigation 2.3: Balanced metre rule

Skills focus

See the Skills grids at the front of this book for details of the skills developed and used in this investigation.

Duration

The practical work will take about 30 minutes; the analysis and evaluation questions will take about 30 minutes.

Preparing for the investigation

- Learners will balance a metre rule using two masses.
- Learners will vary one of the masses and balance the rule again by adjusting its point of suspension from a string loop.
- Learners will determine one of the masses, which is unknown, using their graph.
- Learners will consider limitations of the procedure and suggest improvements.
- This practical can be conducted at the start of a course and does not have to coincide with the teaching of the theory.

Equipment

Each learner or group will need:

- stand
- boss
- clamp
- metre rule
- loop of thick string of circumference 20 cm
- 50 g slotted mass
- three 10 g slotted masses
- sphere of modelling clay (e.g. Plasticine)® of mass 40 g
- small triangular pivot.

Safety considerations

- Learners should take care that the masses and modelling clay do not fall off the metre rule when the rule slides through the string loop.

Carrying out the investigation

When discussing limitations and improvements, explain why the improvement in Table 2.7 is not appropriate (because it could have been achieved with the existing apparatus).

Limitation	Improvement
The slot in a mass means that its centre of mass is not at its centre.	Turn the masses so that the slot is at right angles to the length of the metre rule.

Table 2.7

Theory predicts that a straight line results from plotting $\frac{1}{y}$ against m. Students could plot $\frac{1}{y}$ against m and use $x = 48$ ($\frac{1}{y} = 0.0208$) to read off the value of M from their graph. Figure 2.2 shows how this graph would appear based on the sample results.

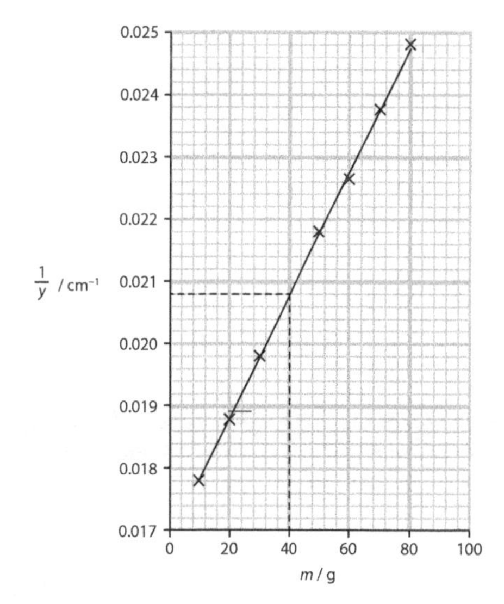

Figure 2.2

Sample results

Table 2.8 provides results the learners may obtain in the investigation.

m / g	y / cm
10	56.1
20	53.1
30	50.5
50	45.9
60	44.2
70	42.1
80	40.3

Table 2.8

Answers to the workbook questions (using the sample results)

a y decreases as m increases.

b, c Graph of y (in cm) against m (in g).

d The value of M is found where $y = 48$ cm. From Figure 2.3, the value of M is 40 g.

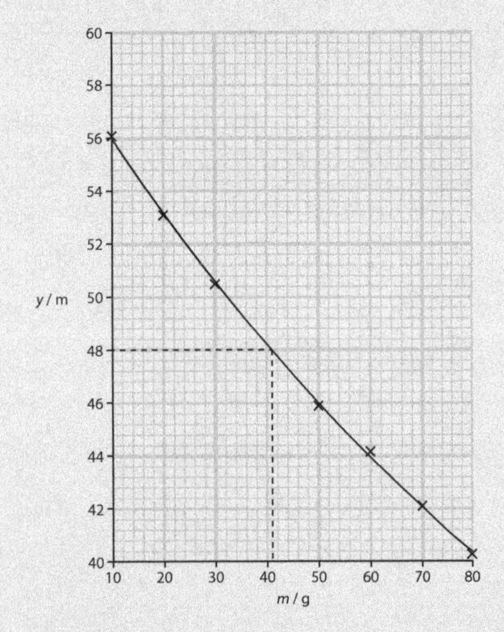

Figure 2.3

	Limitation	Improvement
B	The masses moved on the rule.	Use a small quantity of adhesive putty to keep them in place.
C	The string loop was too small so the rule rotated when balanced.	Use a larger string loop.

Table 2.9

Chapter 3: Kinematics and dynamics

Chapter outline

This chapter relates to Chapter 1: Kinematics, Chapter 2: Motion and Chapter 3: Dynamics, in the coursebook.

In this chapter learners will complete investigations on:

- 3.1 Acceleration of connected masses
- 3.2 Energy and amplitude of a pendulum
- 3.3 Range of a projectile
- 3.4 Terminal velocity of a ball falling through water in a tube.

Practical Investigation 3.1: Acceleration of connected masses

Skills focus

See the Skills grids at the front of this book for details of the skills developed and used in this investigation.

Duration

The practical work will take about 30 minutes and the analysis will take about 30 minutes.

Preparing for the investigation

- If a string over a pulley has a mass attached to each end, any difference in the masses causes the system to accelerate. In this experiment the learner transfers part of the mass from one end to the other in stages. For each stage the motion is timed from rest over a fixed distance.
- In this investigation part of one mass is transferred to the other so that the mass difference is changed but the total mass is constant.
- The mass transferred is a number n of steel washers, so the accelerating force is the weight of $2n$ washers. A graph of acceleration against n should be a straight line.

Equipment

Each learner or group will need:

- pulley wheel to clamp to edge of bench
- thin string
- two mass hangers, each with a total mass of 500 g
- 20 steel washers (steel rings) each of the same approximate mass of 3.3 g (size M10 is suitable)
- two paper clips to act as hooks for the washers
- stopwatch
- metre rule with a millimetre scale
- thick cardboard mat.

See Figure 3.1 in the workbook for the arrangement. The two mass hangers must be able to pass each other.

Safety considerations

- One of the masses will hit the cardboard mat on the floor. The learners must keep their feet away from this area.

Carrying out the investigation

- The measured times are short, so the length of string between its end loops should be as long as possible.
- Learners need to be reminded how to interpret the stopwatch display.

In step **8** the learners will have to develop a reliable technique to start the stopwatch when the upper mass is released and then judge the moment to stop the stopwatch when the mass hits the mat. They should be encouraged to practise before recording their readings.

Given the mass of one washer, more able learners will be able to use their value of gradient from step **e** to deduce a value for the acceleration of gravity.

Common learner misconceptions

- In this experiment both mass hangers together make up the accelerating body. To investigate the relationship between force and acceleration, the mass of the accelerating body is kept constant.

Sample results

$h = 53.9\,\text{cm}$

Learner's results should be similar to those in Table 3.1 (shaded section).

n	t / s				a / cm s^{-2}
	first	second	third	mean	
6	3.35	3.38	3.47	3.40	8.95
8	2.60	2.60	2.66	2.62	15.2
10	2.12	2.15	2.07	2.11	25.2
12	1.82	1.86	1.85	1.84	31.5
14	1.77	1.69	1.73	1.73	36.0
16	1.54	1.60	1.53	1.56	46.1

Table 3.1

Answers to the workbook questions (using the sample results)

a, b See Table 3.1 (unshaded section)

c, d See Figure 3.1.

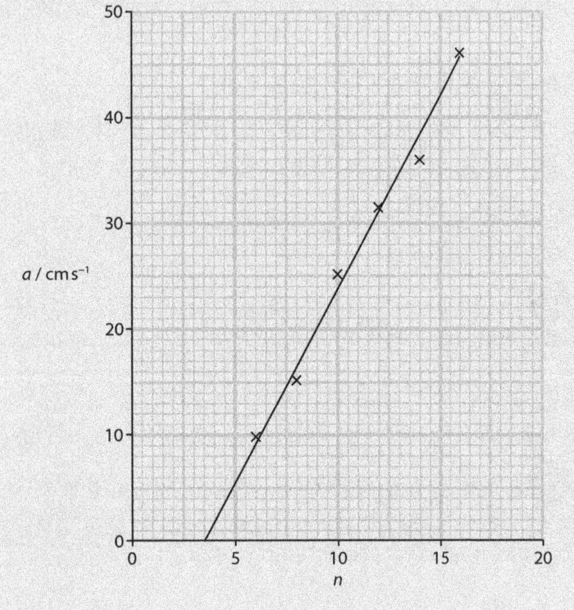

Figure 3.1

e Gradient = 3.61 and intercept = −12.8

f The first washers overcome friction in the pulley. After this more washers can produce an acceleration.

force = mass × a

nmg = mass × a where m = washer mass = 3.3 g

$\frac{a}{n}$ (or gradient) = $\frac{mg}{\text{mass}}$

$g = 1093\,\text{cm s}^{-2}$

Practical Investigation 3.2: Energy and amplitude of a pendulum

Skills focus

See the Skills grids at the front of this book for details of the skills developed and used in this investigation.

Duration

The practical work will take about 30 minutes and the analysis will take about 30 minutes.

Preparing for the investigation

- In this practical exercise the learner sets up a pendulum in its rest position and then hits it with a marble and measures the amplitude of its first swing. This is repeated for different lengths of the pendulum.

Equipment

Each learner or group will need:

- pendulum consisting of a table tennis ball with length of thread attached to it by a spot of glue. The thread should be approximately 70 cm long.
- stand, boss and clamp. The jaws of the clamp must grip the thread securely, so some packing may need to be added to the jaws.
- glass marble of approximate diameter 15 mm in small tray

- inclined rigid pipe held in second stand. A rigid plastic pipe of approximate diameter 2.5 cm and approximate length 30 cm is suitable. It should be clamped securely in a stand with one end 4 cm above the bench and the other end 9 cm above the bench (see Figure 3.2 in the workbook).
- rectangular block with approximate dimensions 15 × 7 × 7 cm. It could be a wooden block or a small brick.
- metre rule.

Safety considerations

There are no special safety issues with this experiment.

Carrying out the investigation

- In step **4** learners will have difficulty in determining *d*. A good technique is to start with the block too far away and then repeat the swing several times, bringing the block a little closer each time until the ball just touches it. *d* can then be measured to the block.

  In step **1** there is an instruction that the inclined pipe must not be adjusted during the experiment. Learners should be told that only the clamp holding the thread can be moved to position the ball at the end of the pipe.

  Values for d^2 in step **a** should be recorded to the same number of significant figures as (or one more than) the significant figures in *d*.

  If more able learners have finished the investigation, suggest they help others who are struggling.

Common learner misconceptions

- Learners may need to be reminded that the length *l* in step **2** is measured to the centre of the ball.

Sample results

Learner's results should be similar to those in Table 3.2 (shaded section).

l / cm	*d* / cm	d^2 / cm^2
56.6	21.7	471
49.8	20.3	412
42.9	18.9	357
37.0	18.2	331
29.3	16.4	269
23.4	15.0	225

Table 3.2

Answers to the workbook questions (using the sample results)

a See Table 3.2 (unshaded section)

b, c See Figure 3.2

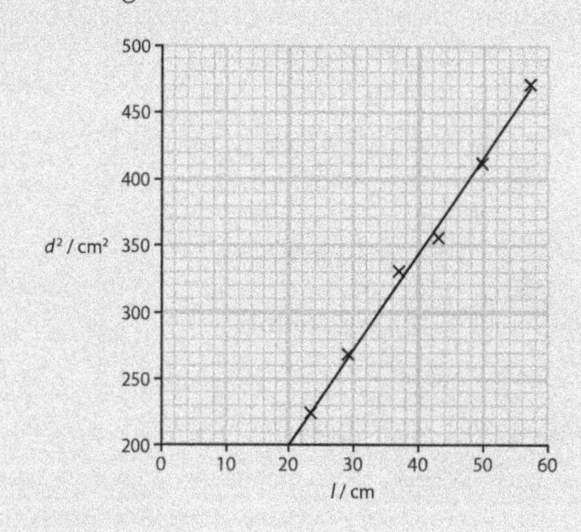

Figure 3.2

d Gradient = 7.2 and intercept = 56

e A = 7.2 cm and B = 56 cm^2

f The marble might be given a velocity at the top of the tube when it is released.

Practical Investigation 3.3: Range of a projectile

Skills focus

See the Skills grids at the front of this book for details of the skills developed and used in this investigation.

Duration

The practical work will take about 30 minutes and the analysis will take about 30 minutes.

Preparing for the investigation

- This investigation uses a launching tube to give a consistent horizontal velocity to a steel ball.
- The launching tube can be made as described in the equipment section, but an equivalent alternative can be used if available.
- Learners launch the ball at different heights above a tray of sand, measuring the horizontal range of the ball for each height.

Equipment

Each learner or group will need:

- curved tube fixed to a rigid cardboard rectangle, as shown in Figure 3.3. The cardboard rectangle should be approximately 20 cm high and 30 cm wide corrugated cardboard: 3 mm thick is suitable. The cut-outs enable the cardboard to be held securely by a stand and clamp. The tube can be flexible plastic tubing with an internal diameter of 8 mm. It can be fixed to the cardboard with glue or tape, and the lower end should be parallel to the edge of the cardboard.

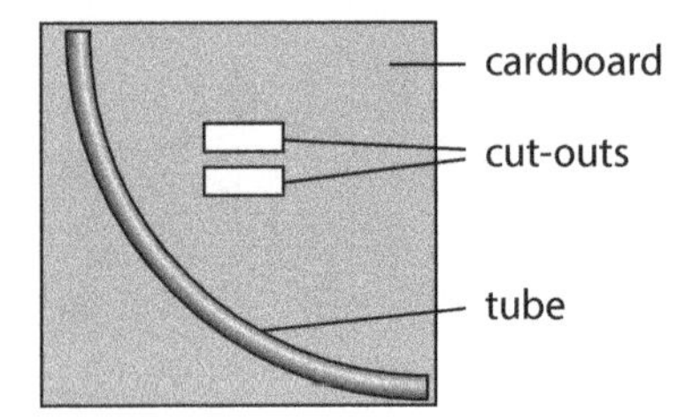

Figure 3.3

- steel ball with approximate diameter 6 mm in a small tray; suitable ball bearings can be obtained from bicycle shops
- tray of sand. The tray should be approximately 30 cm wide and 40 cm long, and the dry sand should be about 2 cm deep.
- pencil
- stand, boss and clamp
- set square
- 30 cm ruler
- metre rule.

Safety considerations

There are no special safety issues with this experiment.

Carrying out the investigation

- In step **2** learners are asked to measure the height h of the tube above the sand. The metre rule is provided for this measurement because its scale starts right at the end of the rule. For measuring D in step **4** the ruler has to be used because the metre rule is too long to lie flat on the sand.
- If the ball rolls after landing, D should be measured to the landing position.

In step **1** learners are asked to lay a pencil on the sand vertically below the end of the tube (this is a marker from which to measure the range of the projectile). It may help to demonstrate how this can be done using the set square and ruler.

In step **7** learners have to reposition the cardboard. It is important that the bottom edge of the cardboard remains parallel to the bench so that the projectile is always launched horizontally.

More able learners could compare the value of v from step **e** with the theoretical launch velocity based on energy transformation in the tube

$$mgx = \frac{1}{2}mv^2$$

where x is the change in height inside the tube.

Common learner misconceptions

- Some learners may be unsure about the independence of the horizontal and vertical components of the ball's velocity. Once the ball has left the tube there is no horizontal force acting on it (if air resistance is ignored) so the horizontal component of its velocity is constant. If it is in the air for longer it will travel further horizontally.

Sample results

Learner's results should be similar to those in Table 3.3 (shaded section).

	D / cm					
h / cm	1	2	3	4	mean	D^2 / cm^2
24.3	31.0	32.0	32.5	30.0	31.4	984
21.6	27.5	29.0	30.5	28.0	28.8	827
17.5	25.7	26.5	25.7	26.8	26.2	685
13.0	19.5	19.7	23.5	23.5	21.6	464
10.0	18.7	21.5	19.0	21.5	20.2	407
6.7	9.7	14.0	18.0	18.5	15.1	227

Table 3.3

 x = 19.3 cm

Answers to the workbook questions (using the sample results)

a See Table 3.3 (unshaded section)

b, c See Figure 3.4

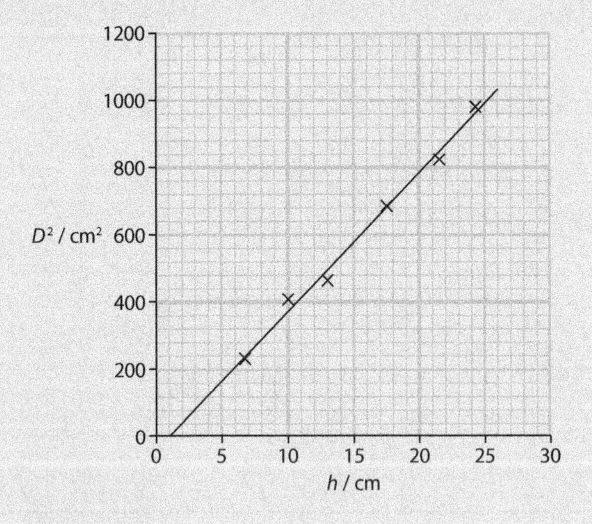

Figure 3.4

d Gradient = 41.4 and intercept = −42.6

e $v = 143\ \text{cm s}^{-1}$

$v = 194\ \text{cm s}^{-1}$

Practical Investigation 3.4: Terminal velocity of a ball falling through water in a tube

Skills focus

See the Skills grids at the front of this book for details of the skills developed and used in this investigation.

Duration

The practical work will take about 15 minutes and the analysis will take about 30 minutes.

Preparing for the investigation

- The investigation does not involve a graph. Instead it measures two sets of results and compares how well they fit a suggested relationship.
- This exercise uses a tall plastic U-tube which must be set up ready for them to carry out the investigation. Each learner or group will only need the U-tube for about 10 minutes so it can be shared between groups.
- The learner drops a steel ball into the tube and times its fall between two marks so that the terminal velocity can be calculated. The internal diameter of the tube is also measured so that the area of the gap between ball and tube can be calculated. This is repeated for the second size of ball.

Equipment

Each learner or group will need:

- a tall U-shaped plastic tube with an internal diameter of 8 mm. It should be filled with water as shown in Figure 3.5. The height of the U-tube should be 1.4 m and the two marks (made with a marker pen) should be 98.5 cm apart. The tray is to catch spilt water and the weight is to keep the tubing in position. The clamps must not crush the tubing: it may be necessary to fix the tubing to them with tape.

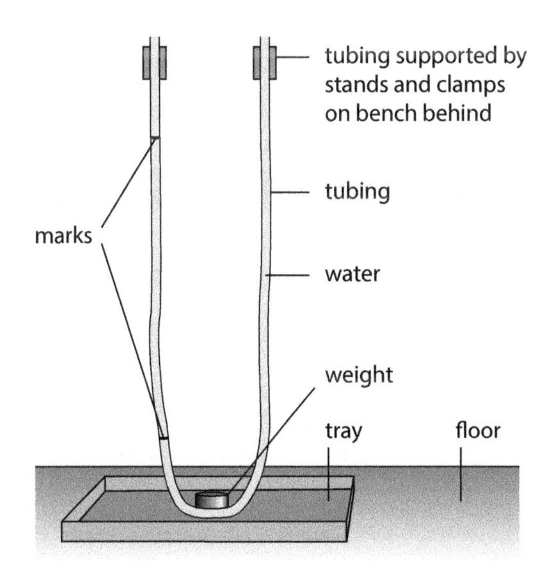

Figure 3.5

- short sample of the same plastic tube
- two sizes of steel ball (five of each size) in a small tray. Ball bearings with diameters of 3.17 mm and 5.53 mm are suitable (both can be obtained from bicycle shops)
- digital calipers
- stopwatch reading to 0.01 s
- bar magnet (for lifting the balls from the tube)
- metre rule.

Safety considerations

There are no special safety issues with this experiment.

Carrying out the investigation

- In step **2** the tubing may not be circular. In this case the value of D must be the average of several diameters.
- The main difficulty will be measuring the time T in step **4** as some values will be less than 2 seconds and the learner has to watch the ball passing each of the marks. They may want to practise first in which case they will have to retrieve the balls using the magnet.

  In step **e** learners may need to be reminded that the percentage difference between the k values is given by $100 \times \frac{(k_1 - k_2)}{\text{average } k}$.

  If there is time, learners could be asked to describe a timing method which would reduce the uncertainty.

Sample results

Learner's results should be similar to those provided here. This data can be used to answer the data analysis questions if learners are unable to do the investigation.

$L = 0.992\,\text{m}$ $D = 8.0\,\text{mm}$

	d / mm	Values of T / s				
Small balls	3.17	1.55	1.57	1.54	1.55	1.54
Large balls	5.53	2.26	2.23	2.25	2.24	2.27

Table 3.4

Answers to the workbook questions (using the sample results)

a, b, c, d and **e**

	T / s	v / m s^{-1}	A / mm^2	k
Smaller balls	1.56	0.641	42.4	0.0151
Larger balls	2.25	0.444	26.2	0.0169

Table 3.5

f Percentage difference = 11.3%

g Percentage uncertainty = $\frac{100 \times 0.2}{1.56}$ = 12.8%

h The variation in k values could be due to variation in the data because the percentage uncertainty in the data is greater than the percentage variation in k.

For an extension, the learner could suggest recording a video of the tube with a timer in view, or using light gates at the measurement positions linked to a timer.

Chapter 4: Forces, work and energy

Chapter outline

This chapter relates to Chapter 4: Forces, Chapter 5: Work, energy and power and Chapter 6: Momentum, in the coursebook.

In this chapter learners will complete investigations on:

- 4.1 Effect of load position on beam supports
- 4.2 Determining the density of a metal sample
- 4.3 Equilibrium of a pivoted wooden strip
- 4.4 Using kinetic energy to do work against friction.

Practical Investigation 4.1: Effect of load position on beam supports

Skills focus

See the Skills grids at the front of this book for details of the skills developed and used in this investigation.

Duration

The practical work will take about 30 minutes and the analysis will take about 20 minutes.

Preparing for the investigation

- A beam is a common structural component. Learners could be asked to think of examples (e.g. floor joist, simple bridge).
- In this practical the learner has to support a beam at both ends and then measure one of the supporting forces as a load is moved along the beam.
- It is a straightforward introductory exercise for this chapter.

Equipment

Each learner or group will need:

- wooden strip with an approximate cross section of 3 cm × 2 cm and a length of 60.0 cm. A hole should be drilled through the strip 1.0 cm from each end. Loops of string of circumference 20 cm should be tied through the holes.
- a separate loop of string of circumference 20 cm
- 0–10 N newton-meter with 0.1 N divisions
- 300 g mass with a hook. Its value should be hidden and it should be labelled *M*
- two stands and two bosses
- metre rule.

Safety considerations

There are no special safety issues with this experiment.

Carrying out the investigation

- Some types of newton-meter have a pointer at some distance from the scale. Learners should be reminded to view the scale perpendicularly to avoid parallax error.
- In step **d** the graph gradient and intercept values should be used from step **c** to find *W* and *Z*. Some learners miss this instruction and calculate *W* and *Z* using two pairs of their readings in simultaneous equations.
- Remind learners that they must carry out step **4** before each set of readings, making sure that the newton-meter is still vertical.

In step **3** of the procedure, learners adjust the height of the boss supporting the newton-meter.

Sample results

The learners' results should be similar to those in Table 4.1.

L = 60.9 cm

x / cm	*F* / N
2.0	1.0
9.1	1.3
18.8	1.8
30.6	2.3
41.8	2.9
58.9	3.7

Table 4.1

Answers to the workbook questions (using the sample results)

a, b See Figure 4.1.

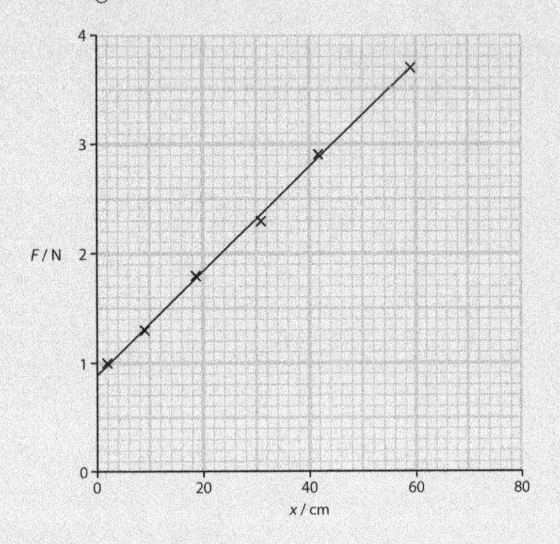

Figure 4.1

c Gradient = 0.047 and intercept = 0.884

d W = 2.86 N Z = 1.77 N

e a top-pan balance

Practical Investigation 4.2: Determining the density of a metal sample

Skills focus

See the Skills grids at the front of this book for details of the skills developed and used in this investigation.

Duration

The practical work will take about 20 minutes and the analysis will take about 20 minutes.

Preparing for the investigation

- In this practical exercise the learner uses a pivoted metre rule to balance a metal sample against a 100 g mass. The process is repeated with the metal sample immersed in water.
- The exercise can reinforce understanding of turning moments and upthrust forces, although prior knowledge is not essential as any necessary relationships are provided in the analysis section.

Equipment

Each learner or group will need:

- metal sample. Any type of metal is suitable. The sample should have an irregular shape and a mass of approximately 100 g. A loop of thread should be tied to it so that it can be suspended from the metre rule (see Figures 4.2 and 4.3 in the workbook).
- wooden metre rule (not plastic as this would bend too much)
- 100 g mass with loop of thread attached so that it can be suspended from the metre rule
- beaker just large enough to immerse the metal sample (see Figure 4.3 in the workbook)
- water in a jug or another beaker
- knife-edge pivot. This can be a wooden or glass triangular prism
- wooden block to support the pivot at a suitable height (see Figures 4.2 and 4.3 in the workbook)
- paper towels to mop up any spillage.

Safety considerations

- There may be some water spillage, but there are no other special safety issues with this experiment.

Carrying out the investigation

- For the second measurement, the metal sample should be completely immersed but not resting on the base of the beaker. To achieve this, the beaker may have to be raised above the bench on a wooden support.
- Some learners spend too much time trying to make the rule perfectly horizontal. This is not necessary: approximately horizontal is acceptable as long as the rule is very nearly stationary.
- When deciding on the significant figures for V in step **b** the learner should consider the significant figures in the value of $(Q - P)$ rather than in Q and P separately.
- In step **c** expect a value for D between about 2 and 8 g cm^{-3}
- If the learner asks for a hint at step **f** suggest thinking about whether the centre of mass of the rule is at the 50 cm mark.

For each reading in step **2**, learners should check that the pivot is at the 50 cm mark.

Some learners will need to be reminded of the need to include a quantity and correct unit for the column headings in all their tables.

Sample results

$P = 21.6\,\text{cm}$ $Q = 17.7\,\text{cm}$

> **Answers to the workbook questions (using the sample results)**
>
> **a** $M = 72.0\,\text{g}$
>
> **b** $V = 13.0\,\text{cm}^3$
>
> **c** $D = 5.54\,\text{g cm}^{-3}$
>
> **d** $D = 5540\,\text{kg m}^{-3}$. A large steel nut was used for the sample results.
>
> **f** If the centre of mass of the metre rule itself is not at the 50 cm mark, the distances P and Q will be incorrect.

Practical Investigation 4.3: Equilibrium of a pivoted wooden strip

Skills focus

See the Skills grids at the front of this book for details of the skills developed and used in this investigation.

Duration

The practical work will take about 30 minutes and the analysis will take about 30 minutes.

Preparing for the investigation

- In this practical the learner has to suspend a wooden strip from one end and then use a spring to apply a horizontal force to the strip.
- The length of the spring is measured rather than the force, so it would be relevant to discuss the relationship between the two (Chapter 7 in the coursebook).
- The exercise is an opportunity for learners to measure angles and plot a graph using a trigonometrical function.

Equipment

Each learner or group will need:

- wooden strip with approximate cross section 3 cm × 2 cm and length 55 cm, with a hole 1.0 cm from one end and another hole at its centre. A string loop of circumference 50 cm should be tied through the central hole and through one of the end loops of the spring (see Figure 4.4 in the workbook).
- Steel tension spring with a loop at each end and a spring constant of approximately $25\,\text{N m}^{-1}$. Examples are Phillip Harris product code B8G87194, which can be found on the Phillip Harris website, and Timstar code SP13863, which can be found on the Timstar website.
- nail to act as pivot. It should be a loose fit in the hole near the end of the rule. The sharp point should be filed off.
- two stands
- two G-clamps
- boss: it should be possible to hold the nail firmly in the boss
- plumb line approximately 30 cm long and with a loop at one end
- 180° protractor with 1° divisions
- metre rule
- ruler
- modelling clay.

Safety considerations

- The horizontal forces may cause the stands to tilt, so their bases should be clamped to the bench.

Carrying out the investigation

- The spring may slide down the stand for low values of θ. It can be secured with a small piece of modelling clay.
- Learners will find the angle measurement difficult as the protractor has to be held steadily in mid-air. However, with care they should be able to measure to the nearest degree.
- Learners have to make sure they measure between the same two points on the coiled section each time.

  Measuring the complete length (including end loops) with a ruler could produce a lot of parallax error; therefore, the coiled section is measured.

  Applying the conditions for equilibrium (Chapter 4 in the coursebook) we get $mg\tan\theta = kL - kL_0$. If m (the mass of the wooden strip) is measured, the gradient of the graph can be used to find the spring constant k.

Sample results

The learners' results should be similar to those in Table 4.2 (shaded section).

θ / °	L / cm	tan θ
18	2.8	0.325
28	3.8	0.532
30	4.3	0.577
33	4.9	0.649
47	7.0	1.072
56	9.2	1.483

Table 4.2

For the extension, mass of wooden strip = 142 g

Answers to the workbook questions (using the sample results)

a See Table 4.2 (unshaded section).

b, c See Figure 4.2.

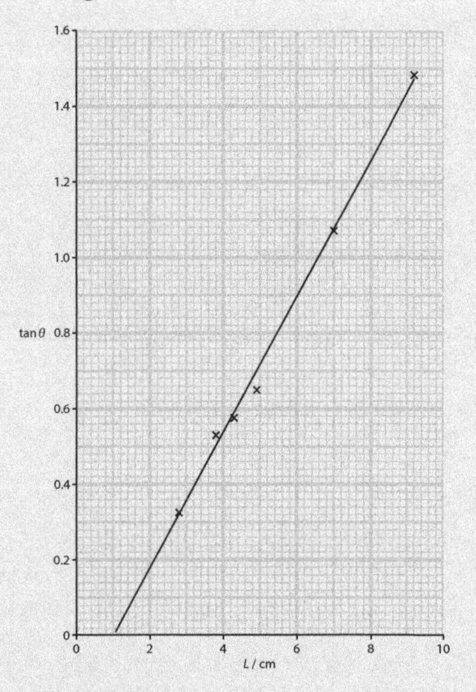

Figure 4.2

d Gradient = 0.180 and intercept = −0.188

e $\tan\theta = 0.180L - 0.188$

f The newton-meter is much heavier than the spring. It would exert an unwanted vertical force on the wooden strip.

For the extension k = gradient × mg

$= 18.0\ \text{m}^{-1} \times 0.142\ \text{kg} \times 9.81\ \text{N kg}^{-1}$

$= 25.1\ \text{N m}^{-1}$

Practical Investigation 4.4: Using kinetic energy to do work against friction

Skills focus

See the Skills grids at the front of this book for details of the skills developed and used in this investigation.

Duration

The practical work will take about 30 minutes and the analysis will take about 30 minutes.

Preparing the investigation

- Learners may not be familiar with the use of piles in civil engineering, so some initial discussion will be helpful.
- The experiment is a simulation of pile-driving and involves using a slide-hammer to drive a ballpoint pen into rice. The slide-hammer is made from a mass hanger and a slotted mass.

Equipment

Each learner or group will need:

- tall plastic container filled with uncooked, dry rice. The container should be approximately 25 cm tall and 3 cm diameter, e.g. a plastic measuring cylinder.
- ballpoint pen of the type shown in Figure 4.3. Use a fine permanent black pen to make seven marks on the parallel section of the barrel. The first mark should be 5 cm from the start of the parallel section and should be labelled 5. The rest of the marks should also be labelled with their distance from the start of the parallel section.

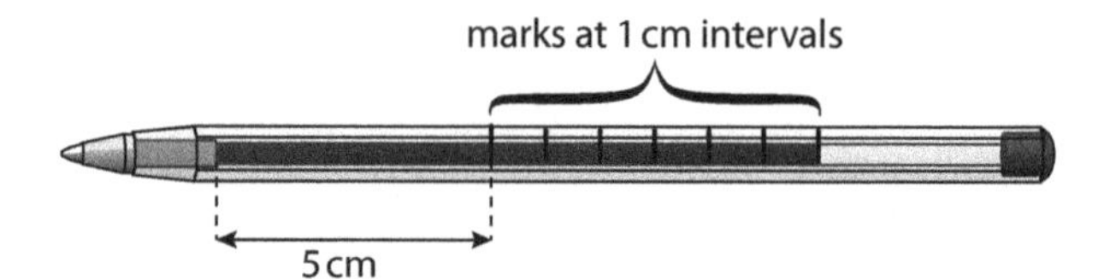

Figure 4.3

- 50 g mass hanger with a 50 g slotted mass. The slotted mass should slide freely on the wire of the hanger. With the slotted mass on the hanger, wind adhesive tape around the wire section 5 cm above it to act as a stop to limit the movement. See Figure 4.6 in the workbook.
- 30 cm ruler.

Safety considerations

- A glass container is not suitable because the hammering action might break it.

Carrying out the investigation

- Learners will probably have to practise keeping the pen vertical as it is hammered into the rice. If they feel that they need to start again they should empty the rice from the container and then put it back (they will need an additional container such as a beaker if they do this).
- Some learners may record the total number of impacts instead of the number for each increment of depth.
- The measurement in step **6** is difficult to make with a ruler. Depending on the shapes of mass and hanger, it may be easier to measure how far the base of the slotted mass can move.

  Learners will need to follow the instructions in step **5** carefully, ensuring that they record the number of impacts needed for the pen to move from one marker to the next.

  If they have time, learners could be asked to suggest a change to the experiment to overcome the difficulty they have described in step **e**.

Sample results

The learners' results should be similar to those in Table 4.3 (shaded section).

D / cm	n	F / N
5.0	6	13.2
6.0	10	22.1
7.0	13	28.7
8.0	12	26.5
9.0	15	33.1
10.0	18	39.7

Table 4.3

$h = 0.045$ m

Answers to the workbook questions (using the sample results)

a $E = 0.0221$ J

b See Table 4.3 (unshaded section).

c See Figure 4.4.

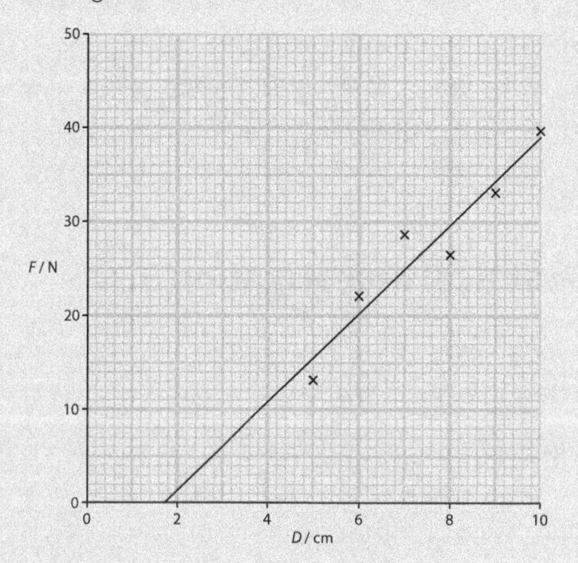

Figure 4.4

d The positive trend shows that the friction force increases as the length of the pen below the surface increases.

The friction force varies with the area of the pen in contact with the rice.

e It is difficult to judge when the next mark reaches the surface of the rice because the surface of the rice is uneven.

or The depth of the pen is not accurate because the pointed part of the pen is not considered.

or Other sensible comment with reason.

A circle of card around the pencil could rest on the surface of the rice to give a definite level.

or The total area of the pen surface in contact with the rice (including the conical part and the cylindrical part) could be plotted instead of D.

Chapter 5: Matter and materials

Chapter outline

This chapter relates to Chapter 7: Matter and materials, in the coursebook.

In this chapter learners will complete investigations on:

- 5.1 Finding the Young modulus for nylon
- 5.2 Using a spring to find the Young modulus for steel
- 5.3 Water pressure and flow rate.

Practical Investigation 5.1: Finding the Young modulus for nylon

Skills focus

See the Skills grids at the front of this book for details of the skills developed and used in this investigation.

Duration

The practical work will take about 30 minutes and the analysis will take about 30 minutes.

Preparing for the investigation

- In this practical exercise the learner has to measure a marked section of nylon thread and then monitor its extension as the tension in the thread is increased.
- The data is used to find a value for the Young modulus.
- A similar arrangement of apparatus could be used to find the Young modulus for a metal in the form of a wire.

Equipment

Each learner or group will need:

- 0.15 mm diameter nylon thread with loops. It should have a single filament of nylon (e.g. fishing line). Tie a loop at each end using the knot shown in Figure 5.1 to prevent slipping. The total length including the loops should be approximately 1.2 m.

Figure 5.1

- pulley suitable for clamping to the table top. If a pulley is not available a short length of smooth metal pipe could be fixed along the edge of the table top.
- wooden block approximately 10 cm by 10 cm by 10 cm. It should have a hook screwed into one of the vertical faces at the same height as the top of the pulley.
- G-clamp to clamp the wooden block to the table top
- mass hanger
- five 100 g slotted masses to fit on the mass hanger
- metre rule with a millimetre scale
- sticky tape for the learner to attach to the nylon thread to mark a section
- scissors to cut the sticky tape
- micrometer.

Safety considerations

- Although the breaking tension for 0.15 mm diameter nylon thread is approximately 20 N, the nylon thread may break during testing, releasing stored elastic energy.
- Learners should be instructed to wear safety goggles whenever the nylon thread is under tension.

Carrying out the investigation

- The changes in length are small and so the pointer readings must be made carefully. If the pointers are not touching the metre rule they should be gently pushed down so that they do touch it when readings are taken.

These notes apply to the sections of the procedure described in the workbook:

- To make a pointer (step **1**) a length of sticky tape should be folded over the nylon thread and the end cut to make a pointer touching the metre rule scale. The strip of tape must be at right angles to the nylon thread.
- In step **d** the learner should choose scales that enable coordinates to be read easily, and that make the plotted points occupy at least half the grid in the vertical and horizontal directions.
- The graph will probably be linear for the range of masses used. If it curves at higher loads the learner can choose to use the gradient of the lower, more linear section at step **e**.
- At step **5** the measurements made with a micrometer should be recorded to the nearest 0.01 mm; for example, 0.30 mm or 0.17 mm. The average of several values can be recorded with one extra significant figure; for example, 0.185 mm.
- At step **i**, if the learner asks for a hint, suggest thinking about the dimensions of the wire (thicker or thinner, longer or shorter).

  The more able learners could, given that the Young modulus of steel is 50 times that of nylon, estimate the diameter of a steel wire that would give the same extensions in this experiment.

  Some learners will need to be reminded of the need to include a quantity and correct unit for the column headings in all their tables.

Common learner misconceptions

- This experiment starts with the mass hanger already attached to keep the nylon thread straight so that the initial length between pointers can be measured.
- Learners may be concerned that the mass of the hanger itself is not included when the force is calculated, but only the changes in force and length are needed to determine the Young modulus.

Sample results

Learner's results should be similar to those shown below and in Table 5.1 (shaded section).

y_1 = 11.9 cm y_2 = 80.7 cm (at step **1**)

Total added mass *M* / kg	y_1 / cm	y_2 / cm	*F* / N	*x* / cm
0.100	12.0	82.1	0.981	1.3
0.200	12.2	83.7	1.96	2.7
0.300	12.4	84.5	2.94	3.3
0.400	12.5	85.6	3.92	4.3
0.500	12.6	86.2	4.91	4.8

Table 5.1

Answers to the workbook questions (using the sample results)

a L = 68.8 cm

b, c See Table 5.1 (unshaded section)

d, e See Figure 5.2

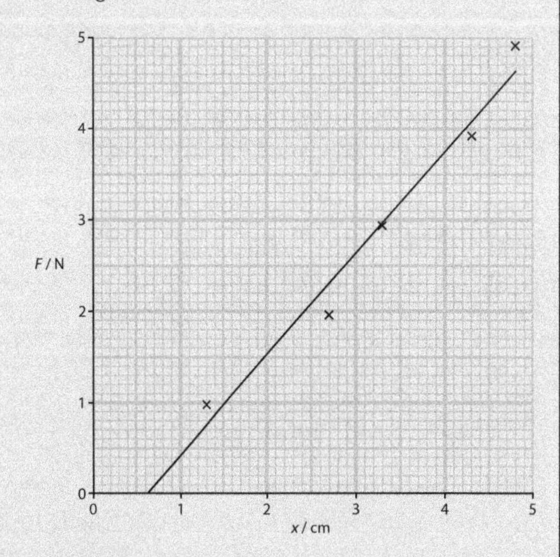

Figure 5.2

f Gradient = 1.11

g $k = 111\ \text{N m}^{-1}$

5 Average d = 0.15 mm

h L = 0.688 m d = 0.000 15 m $E = 4.32 \times 10^9\ \text{N m}^{-2}$

i The steel wire could have smaller diameter than the nylon thread

or the steel wire could be longer than the nylon thread

or larger masses could be used.

The diameter would have to be reduced by a factor of $\sqrt{50}$.

Practical Investigation 5.2: Using a spring to find the Young modulus for steel

Skills focus

See the Skills grids at the front of this book for details of the skills developed and used in this investigation.

Duration

The practical work will take about 20 minutes and the analysis will take about 20 minutes.

Preparing for the investigation

- In this practical the learner has to measure the dimensions of a steel spring and then find the period of its oscillations in a mass–spring system.
- The data is used to find a value for the Young modulus of steel.
- It is an indirect method of finding the Young modulus, but it provides an opportunity for learners to time oscillations.

Equipment

Each learner or group will need:

- steel tension spring with a loop at each end and a spring constant of approximately $25\,N\,m^{-1}$
- digital calipers measuring to 0.01 mm. If only a micrometer is available, the learner could be given a single coil cut from an identical spring to enable an accurate measurement of the wire diameter to be made.
- 300 g mass with a hook. This could be a 100 g mass hanger with two 100 g slotted masses taped onto it.
- stand, boss and clamp
- stopwatch reading to 0.01 s or better.

Safety considerations

There are no special safety issues with this experiment.

Carrying out the investigation

- In step **1** to find the diameter of the wire used to make the spring the learner needs to measure a flat section of the spring. Bending the spring opens up the coils and gives access for the jaws of the callipers.

In step **1** the end loops should not be included in n.

In step **b** Young modulus for steel is $210 \times 10^9\,N\,m^{-2}$ but learners may arrive at a lower value. If so, it will probably be caused by zinc plating on the wire. This means that the steel diameter is **less** than the measured diameter.

A learner who finishes quickly could be asked how many turns an otherwise identical spring would have to have to give double the time period.

Common learner misconceptions

- In step **4** some learners start timing from the moment the mass is released, but it is better to watch for a while before choosing the point in the oscillation at which counting will be done.
- A possible error is to start counting at one instead of zero, so that the time is measured for nine oscillations instead of ten. Another possible error is to count half oscillations instead of whole ones.

Sample results

Learner's results should be similar to those shown below and in Table 5.2 (shaded section).

d = 0.68 mm, 0.68 mm, 0.68 mm

Average d = 0.68 mm

D = 16.08 mm, 16.04 mm, 16.07 mm

Average D = 16.06 mm

n = 28

$10T$ / s			Mean $10T$ / s	T / s
6.83	6.84	6.82	6.83	0.683

Table 5.2

Answers to the workbook questions (using the sample results)

a See Table 5.2 (unshaded section)

b $E = 1.71 \times 10^{11}\,N\,m^{-2}$. Should be to 3 significant figures (as shown).

c E = 171 GPa

d 0.1 mm

e *d* would change from 0.68 mm to 0.7 mm, and the % uncertainty would be 14%.

n would have to be increased by a factor of 4.

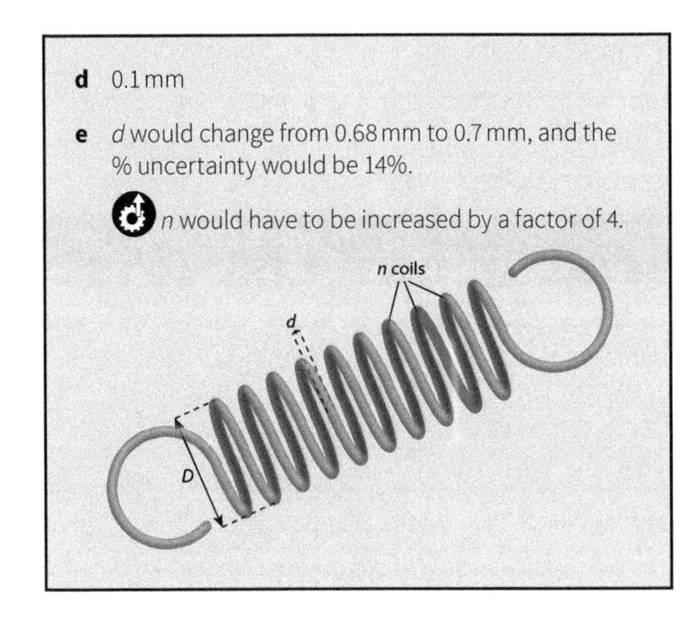

Practical Investigation 5.3: Water pressure and flow rate

Skills focus

See the Skills grids at the front of this book for details of the skills developed and used in this investigation.

Duration

The practical work will take about 30 minutes and the analysis will take about 30 minutes.

Preparing for the investigation

- In this practical the learner has to determine the cross-sectional area of a container and measure the rate of change of water level as the water escapes through a hole in the base.
- The data is used to test whether the flow rate is proportional to the water pressure at the hole.

Equipment

Each learner or group will need:

- a two litre clear plastic water bottle with a constant circular cross section. A 2.5 mm diameter hole should be drilled in one of the lowest parts of the base near to the edge. The top of the bottle should be cut off to give a container 20 cm tall with the shape shown in Figure 5.3.

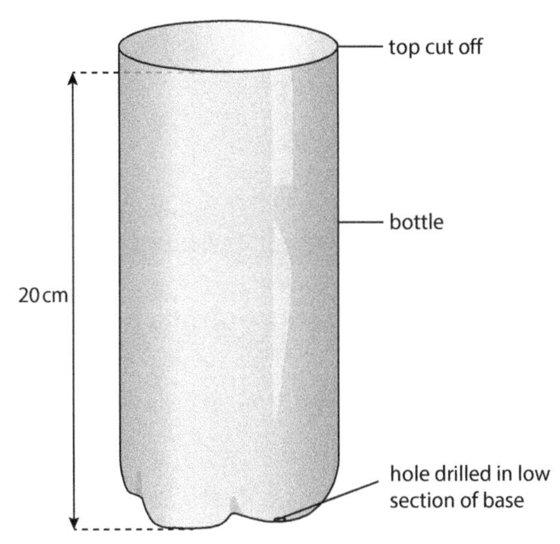

Figure 5.3

- flat, horizontal worktop with sink to collect water: see Figure 5.4 in the workbook. If this is not available a bowl can be positioned below the edge of a table.
- two-litre capacity jug or beaker
- water supply
- stopwatch with lap timer feature
- stand, boss and clamp
- metre rule.

Safety considerations

There are no special safety issues with this experiment.

Carrying out the investigation

- The top of the container will probably be very flexible and difficult to measure with a metre rule, but learners could help each other at this stage: there will be a chance to comment on this later.

  At step **4** the learner must fill the container so that the water level is higher than the starting point, then wait until the level falls to 18 cm before starting the stopwatch.

  Practice using the lap timer will be needed before recording any times at step **5**. The only alternative to lap times is to time from the start down to each of the levels separately: this would mean refilling the container a total of 8 times.

In part **e** the gradient has a negative value because the volume left in the container gets less with time. The minus sign must disappear when the flow rate out of the container is calculated.

Learners who finish before the others could draw a third tangent to the curve and calculate a third value for k.

Common learner misconceptions

- Each timed interval is not the time between levels, it is the time from the start (at 18 cm) to that level.

Sample results

Learner's results should be similar to those shown below and in Table 5.3.

d = 11.1 cm, 11.7 cm, 11.4 cm

Water level height h / cm	Time to fall from 18 cm to h T / s
18.0	0
16.0	27
14.0	55
12.0	85
10.0	118
8.0	154
6.0	196
4.0	249
2.0	314

Table 5.3

Answers to the workbook questions (using the sample results)

a $A = \dfrac{\pi \times 11.4^2}{4} = 102.1\ \text{cm}^2$

b, c See Figure 5.4

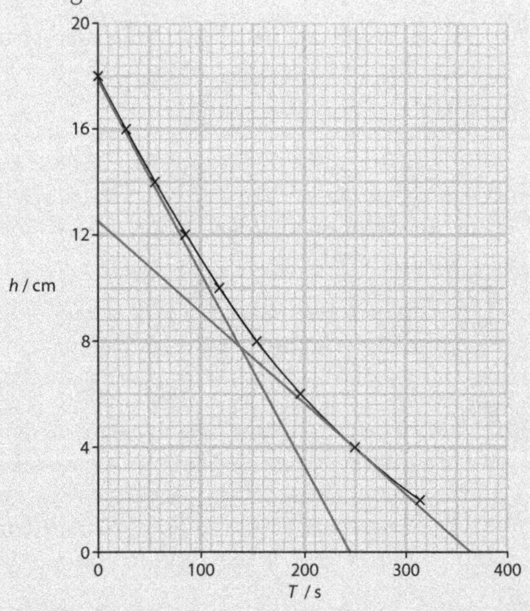

Figure 5.4

d Gradient = −0.072

e $F = 7.35\ \text{cm}^3\,\text{s}^{-1}$

f $P = 1570\ \text{N m}^{-2}$

The unit for h must be changed from cm to m before calculating the pressure.

g Gradient = − 0.035, $F = 3.57\ \text{cm}^3\,\text{s}^{-1}$ and $P = 392\ \text{N m}^{-2}$

h First $k = 0.185$ and second $k = 0.180$

The main source of uncertainty in k is the drawing of the tangent. This uncertainty could be roughly estimated at 10%, so if the difference between the two k values is more than 10% the hypothesis is not supported by the data. These k values are only 3% apart, so they support the hypothesis.

i The flexibility of the plastic container means it distorts when touched, so it is difficult to measure the diameter.

If viewed from above, a photo with a scale in view could be used.

j The video would have to show the water level, the metre rule and a clock.

A third tangent drawn at $h = 8.0$ cm gives a k value of 0.200 (still within 10% of the other values).

Chapter 6:
Electric current, potential difference and resistance

Chapter outline

This chapter relates to Chapter 9: Electric current, potential difference and resistance and Chapter 10: Kirchhoff's laws, in the coursebook.

In this chapter learners will complete investigations on:

- 6.1 Power and resistance of a lamp
- 6.2 Resistors in series
- 6.3 Resistors in parallel.

Practical Investigation 6.1:
Power and resistance of a lamp

Skills focus

See the Skills grids at the front of this book for details of the skills developed and used in this investigation.

Duration

The practical work will take about 30 minutes; the analysis and evaluation questions will take about 30 minutes.

Preparing for the investigation

- Learners should know the equations linking current, potential difference, resistance and power.
- Learners should be able to set up a circuit from a circuit diagram.
- Learners should be able to plot a curve of best fit through plotted points.
- Learners will measure the potential difference across and the current in a lamp.
- Learners will calculate resistance and power and investigate how they vary with potential difference.

Equipment

Each learner or group will need:

- low resistance power supply variable up to 12 V
- 6 V 60 mA lamp with holder
- two digital multimeters
- five connecting leads
- switch.

Safety considerations

- Learners should be encouraged to switch the circuit off between readings either using the circuit switch or the switch on the power supply.
- Learners should **not** apply a voltage of more than 6 V across the lamp.

Carrying out the investigation

- Learners may get negative readings on their meters because the meters are connected into the circuit the wrong way round.
- If the power supply has several outputs, advise learners which outputs give a variable voltage of up to 6 V.

If a 6 V 60 mA lamp is not available and an alternative such as 2.5 V 0.3 W or 12 V 24 W is used, the values of R and P on the axes of the graphs should be changed as shown in Table 6.1.

	R–V graph		*P–V* graph	
Lamp	***y*-axis**	***x*-axis**	***y*-axis**	***x*-axis**
2.5 V 0.3 W	0–25 Ω	0–2.5 V	0–0.3 W	0–2.5 V
12 V 24 W	0–6 Ω	0–12 V	0–24 W	0–12 V

Table 6.1

Some learners will need to be reminded of the need to include a quantity and correct unit for the column headings in all their tables.

If more able learners have finished the investigation, suggest they help other learners who may be struggling.

Sample results

The data in Table 6.2 gives an example of the results the learners should obtain from the investigation.

V / V	I / mA	I / A	R / Ω	P / W
1.00	24.2	0.0242	41.3	0.0242
2.00	33.4	0.0334	59.9	0.0668
3.00	41.5	0.0415	72.3	0.125
4.00	48.9	0.0489	81.8	0.196
5.00	55.4	0.0554	90.3	0.277
6.00	61.2	0.0612	98.0	0.367

Table 6.2

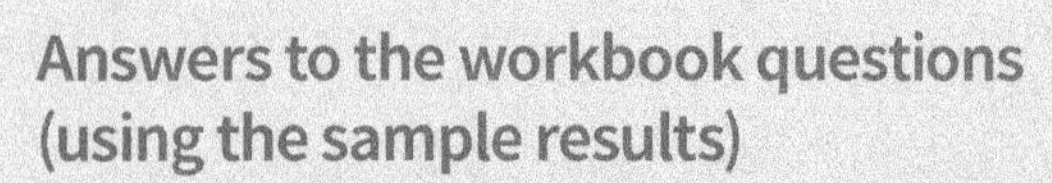

Answers to the workbook questions (using the sample results)

a See the values in Table 6.1.

b, c See Figure 6.1.

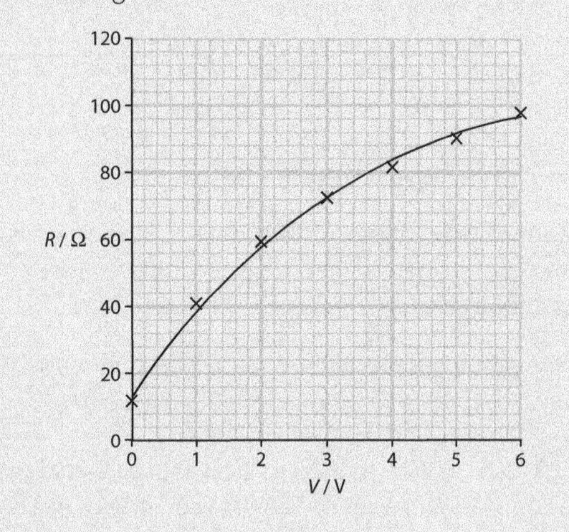

Figure 6.1

d, e See Figure 6.2.

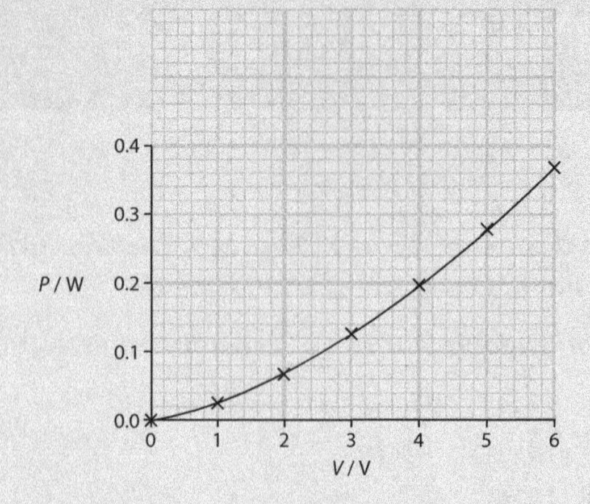

Figure 6.2

f R increases with V. The rate of increase decreases as V increases.

g P increases with V. The rate of increase increases as V increases.

h The two graphs should have the same shape as the plotted ones. The y-axis of the R–V graph should go up to 600 Ω. The y-axis of the P–V graph should go up to 100 W.

i The R–V graph should not go through the point (0, 0) because when $V = 0$ the resistance has a non-zero value.

The P–V graph should go through the point (0, 0) because when $V = 0$ there is no current in the resistor so

$$P = VI = 0$$

Practical Investigation 6.2: Resistors in series

Skills focus

See the Skills grids at the front of this book for details of the skills developed and used in this investigation.

Duration

The practical work will take about 30 minutes; the analysis and evaluation questions will take about 30 minutes.

Preparing for the investigation

- Learners should know the formula for resistors in series.
- Learners should be able to set up a circuit from a circuit diagram.

- Learners should be able to draw graphs.
- Learners should know the equation of a straight line: $y = mx + c$.
- Learners will connect a circuit with two resistors in series and investigate how the current in the circuit changes when the value of one of the resistors is changed.

Equipment

Each learner or group will need:

- 1.5 V cell with terminals
- switch
- five connecting leads
- two component holders suitable for resistors
- digital multimeter with a 0–200 mA scale
- resistors clearly labelled with the following values: 18 Ω, 22 Ω, 27 Ω, 33 Ω and 2 × 15 Ω.

Safety considerations

- Learners should be encouraged to switch the circuit off between readings using the circuit switch.

Carrying out the investigation

- Learners may get a negative reading on the meter because it is connected into the circuit the wrong way round.

Some learners will need to be reminded of the need to include a quantity and correct unit for the column headings in all their tables.

If more able learners have finished the investigation, suggest they help other learners who may be struggling.

Sample results

These should give an idea of the results the learners should end the investigation with.

$E = 1.49\,\text{V}$

R / Ω	I / mA	$\frac{1}{I}$ / A^{-1}
15	45.7	21.9
18	41.9	23.9
22	37.4	26.7
27	33.5	29.9
33	29.4	34.0

Table 6.3

Answers to the workbook questions (using the sample results)

a I decreases as R increases

b, c and **d** See Figure 6.3.

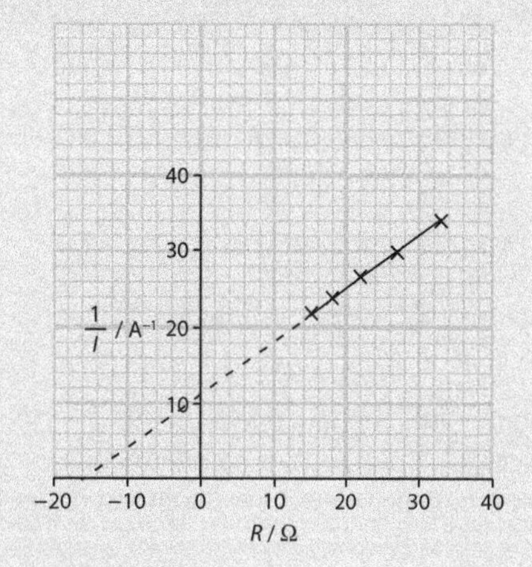

Figure 6.3

e 0.672

f $E = \frac{1}{0.672} = 1.49\,\text{V}$

g $11.8 = \frac{X}{1.49}$

h $X = 11.8 \times 1.49 = 17.6\,\Omega$

i −17.6

j $X = 17.6\,\Omega$

k, l

	Smaller	Same	Bigger
Gradient		✓	
y-intercept	✓		
x-intercept	✓		

Table 6.4

m $E = 1.49\,\text{V}$ so percentage difference = 0%

n Plotted points are too far away from the x-intercept.

Practical Investigation 6.3: Resistors in parallel

Skills focus

See the Skills grids at the front of this book for details of the skills developed and used in this investigation.

Duration

The practical work will take 30 minutes; the analysis and evaluation questions will take 30 minutes.

Preparing for the investigation

- Learners should know the formula for resistors in series.
- Learners should be able to set up a circuit from a circuit diagram.
- Learners should be able to draw graphs.
- Learners should know the equation of a straight line: $y = mx + c$.
- Learners will connect a circuit with two resistors in parallel and investigate how the current in the circuit changes when the value of one of the resistors is changed.

Equipment

Each learner or group will need:

- 1.5 V cell with terminals
- switch
- six connecting leads
- two component holders suitable for resistors
- digital multimeter with a 0–200 mA scale
- resistors clearly labelled with the following values: 120 Ω, 150 Ω, 180 Ω, 220 Ω and 2 × 100 Ω.

Safety considerations

- Learners should be encouraged to switch the circuit off between readings using the circuit switch.

Carrying out the investigation

- Learners may get a negative reading on the meter because it is connected into the circuit the wrong way round.

  If more able learners have finished the investigation, suggest they help other learners who may be struggling.

Sample results

These should give an idea of the results the learners should end the investigation with.

$E = 1.49$ V

R / Ω	I / mA	I / A	IR / V
100	28.3	0.0283	2.83
120	26.2	0.0262	3.14
150	23.8	0.0238	3.57
180	22.3	0.0223	4.01
220	20.9	0.0209	4.60

Table 6.5

Answers to the workbook questions (using the sample results)

a I decreases as R increases.

b See values of IR in Table 6.5.

c, d See Figure 6.4.

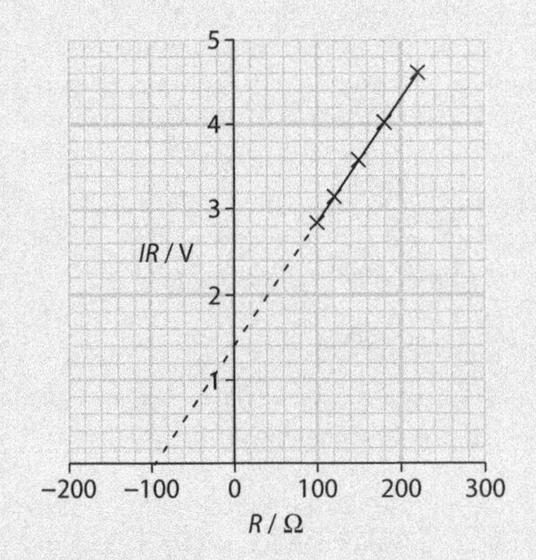

Figure 6.4

e Gradient = 0.0147

f y-intercept = E = 1.37 V

g Gradient = $0.0147 = \frac{1.37}{X}$ so $X = 93\ \Omega$

h $X = -93\ \Omega$

i

	Smaller	Same	Bigger
Gradient			✓
y-intercept		✓	
x-intercept	✓		

Table 6.6

j Line drawn with bigger gradient and same y-intercept

k Percentage difference = $\left[\dfrac{1.49 - 1.37}{\frac{1.49 + 1.37}{2}}\right] \times 100 = 8.4\%$

l Plotted points are too far away from the x-intercept.

Chapter 7: Resistance and resistivity

Chapter outline

This chapter relates to Chapter 11: Resistance and resistivity and Chapter 12: Practical circuits, in the coursebook.

In this chapter learners will complete investigations on:

- 7.1 Resistivity of the metal of a wire
- 7.2 Internal resistance of a dry cell
- 7.3 Potential divider.

Practical Investigation 7.1: Resistivity of the metal of a wire

Skills focus

See the Skills grids at the front of this book for details of the skills developed and used in this investigation.

Duration

The practical work will take about 30 minutes; the analysis and evaluation questions will take about 30 minutes.

Preparing for the investigation

- Learners should be able to recall and use the equation $R = \frac{\rho l}{A}$.
- Learners should be able to draw graphs.
- Learners will investigate how the resistance of a resistance wire varies with length.
- Learners will use a micrometer to measure the diameter of the wire.

Equipment

Each learner or group will need:

- two connecting leads
- two crocodile clips
- digital multimeter
- metre rule.

Access to:

- micrometer
- reel of resistance wire (36 swg constantan diameter 0.19 mm or 32 swg nichrome diameter 0.27 mm)
- adhesive tape
- scissors
- wire cutters.

Safety considerations

There are no particular safety issues with this investigation

Carrying out the investigation

- If the wire is not straight a length of 50 cm read from one end may not be exactly 50 cm of wire because the wire has kinks and bows out a bit.

  Learners may be unsure which range to use on the multimeter (see Investigation 1.3 Part 1).

Sample results

The data in Table 7.1 gives an idea of the results the learners should obtain from the investigation.

l / m	R / Ω
0.100	4.3
0.250	6.5
0.400	9.1
0.550	11.8
0.700	14.4
0.850	17.1

Table 7.1

Diameter d = 0.19 mm

Total resistance of the connecting leads = 1.1 Ω

Answers to the workbook questions (using sample results)

a $A = 0.028\,mm^2 = 2.8 \times 10^{-8}\,m^2$

b, c See Figure 7.1

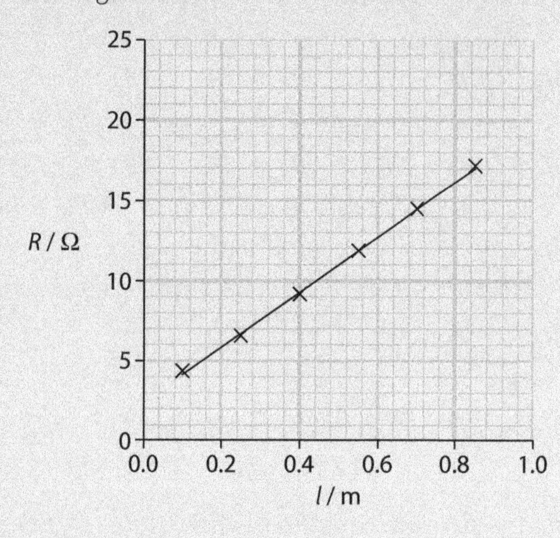

Figure 7.1

d 17.2

e 2.35

f $17.2 \times 2.8 \times 10^{-8} = 4.82 \times 10^{-7}\,\Omega\,m$

g Circle constantan and 0.19

h $1.2 \times 10^{-6} = 17.2 \times A \quad A = 7.0 \times 10^{-8} \quad d = 0.30\,mm$ (choose 0.27 mm or 0.32 mm)

i If $\rho = 1.7 \times 10^{-8}$, $A = 2.8 \times 10^{-8}$ and $l = 1.000\,m$
$R \approx 6 \times 10^{-6}\,\Omega$ (too small) *or* $17.2 \times A = 1.7 \times 10^{-8}$
$d = 36\,km$ (too thick)

j y-intercept and total resistance of connecting leads have similar magnitude.

Practical Investigation 7.2: Internal resistance of a dry cell

Skills focus

See the Skills grids at the front of this book for details of the skills developed and used in this investigation.

Duration

The practical work will take about 30 minutes; the analysis and evaluation questions will take about 30 minutes.

Preparing for the investigation

- Learners should be able to recall and use the equation $V = E - Ir$.
- Learners should be able to set up a circuit from a circuit diagram.
- Learners should be able to draw graphs.
- Learners should know the equation of a straight line: $y = mx + c$.
- Learners will investigate the change in potential difference across a dry cell with the current being drawn from it.

Equipment

Each learner or group will need:

- 1.5 V cell with terminals
- switch
- six connecting leads
- digital multimeter with a 0–200 mA scale reading to the nearest 0.1 mA
- digital multimeter with a 0–2 V scale reading to the nearest 0.001 V
- rheostat.

Safety considerations

- Learners should be encouraged to switch off between readings using the circuit switch.

Carrying out the investigation

- Learners may get a negative reading on the meter because it is connected into the circuit the wrong way round.
- The reading on the voltmeter may fluctuate because the contact formed by the crocodile clips is not perfect. Consider ways of improving contact such as placing a small nail in the rule at the 5 cm mark, soldering the resistance wire to the nail, and connecting one of the crocodile clips to the nail. The length *l* is then measured from the nail and only one crocodile clip has to be attached to the resistance wire.

Sample results

The data in Table 7.2 should give an idea of the results the learners should end the investigation with.

$E = 1.495\,\text{V}$

Maximum voltmeter reading with switch open / V	I / A	V / V
1.493	0.1926	1.442
1.492	0.1740	1.443
1.490	0.1549	1.449
1.490	0.1354	1.454
1.490	0.1189	1.456
1.490	0.1075	1.460
1.490	0.0995	1.462

Table 7.2

Final voltmeter reading = 1.494 V

Resistance of connecting leads / Ω					
0.4	0.4	0.7	0.4	0.5	0.4

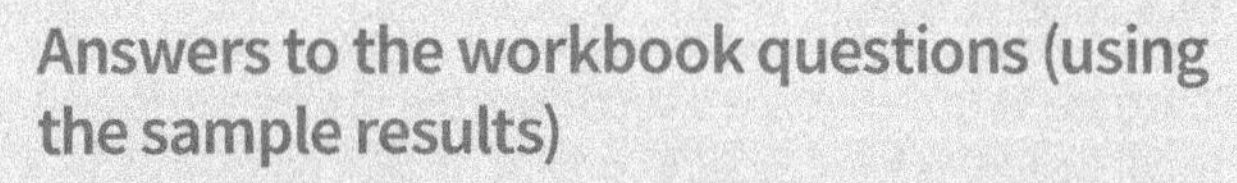

Answers to the workbook questions (using the sample results)

a V decreases as I increases.

b, c See Figure 7.2.

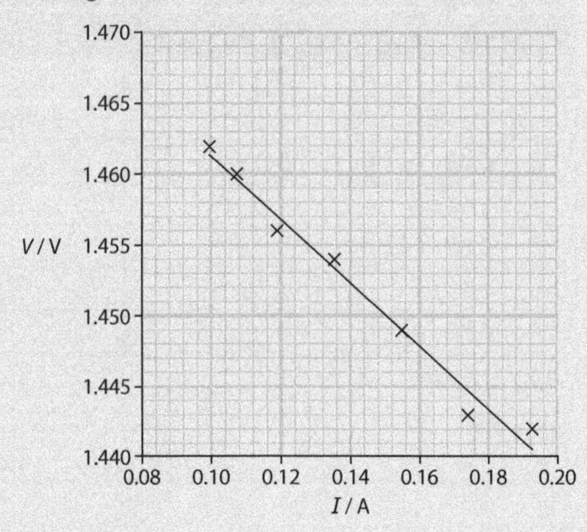

Figure 7.2

d −0.216

e $r = 0.216\,\Omega$

f 1.483

g $E = 1.483\,\text{V}$

h The initial value $E = 1.495\,\text{V}$, the final reading shows a drop to 1.493 V. The value obtained from the *y*-intercept is 1.483 V but this is less reliable because the plotted points show a slight scatter and the intercept is a long way from the plotted points.

All the evidence suggests that the e.m.f. of the cell falls during the experiment but shows some recovery when current is no longer being drawn from it.

i The total resistance of the connecting leads is 2.8 Ω which is greater than the internal resistance of the cell (0.216 Ω)

j If the switch remains closed, current is drawn from the cell and its e.m.f. drops. If E is no longer constant the relationship $V = -Ir + E$ does not fit the equation of a straight line.

Practical Investigation 7.3: Potential divider

Skills focus

See the Skills grids at the front of this book for details of the skills developed and used in this investigation.

Duration

The practical work will take about 30 minutes; the analysis and evaluation questions will take about 30 minutes

Preparing for the investigation

- Learners should be able to recall and use the potential divider equation $ER = V\left(\frac{\rho l}{A} + R\right)$.
- Learners should be able to set up a circuit from a circuit diagram.
- Learners should be able to draw graphs.
- Learners should know the equation of a straight line: $y = mx + c$.
- Learners will connect a resistance wire and a fixed resistor in series with a dry cell and record the potential difference across the resistor as the length of the wire is varied.

Equipment

Each learner or group will need:

- the wire on the metre rule used in Practical investigation 7.1
- 1.5 V cell with terminals
- switch
- six connecting leads
- two crocodile clips
- 15 Ω resistor
- component holder for the resistor
- digital multimeter to measure 0–2 V to the nearest 0.001 V.

Safety considerations

- Learners should be encouraged to switch off between readings using the circuit switch.

Carrying out the investigation

- Learners may get a negative reading on the meter because it is connected into the circuit the wrong way round.

  Some learners will need to be reminded of the need to include a quantity and correct unit for the column headings in all their tables.

Sample results

The data in Table 7.3 gives an idea of the results the learners should end the investigation with.

$E = 1.497$ V

l / m	V / V	$\frac{1}{V}$ / V^{-1}
0.200	1.158	0.8636
0.300	1.062	0.9416
0.400	0.990	1.010
0.500	0.915	1.093
0.600	0.853	1.172
0.700	0.798	1.253
0.800	0.750	1.333

Final $E = 1.496$ V

Table 7.3

Answers to the workbook questions (using the sample results)

a V decreases as l increases

b See values of $\frac{1}{V}$ in Table 7.3.

c, d See Figure 7.3.

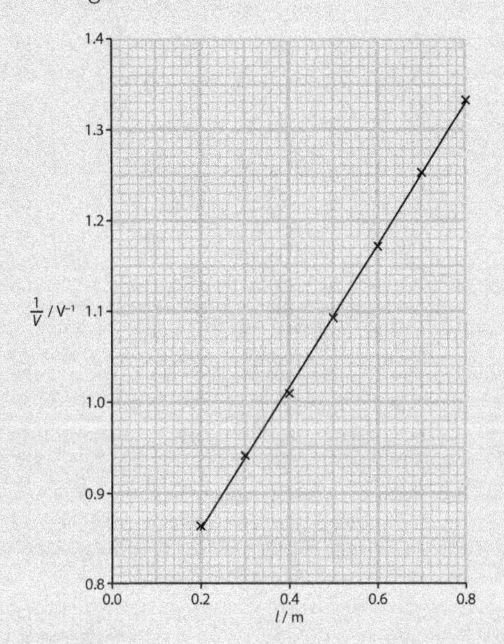

Figure 7.3

e Gradient = 0.784

f y-intercept = 0.703

g $E = 1.422\,V$

h Using $A = 2.82 \times 10^{-7}$, $E = 1.422\,V$, $\rho = 4.82 \times 10^{-7}$ and gradient = 0.784, $R = 15.3\,\Omega$

i The final value of E is below the initial value as expected but not by much. No current was drawn from the cell between readings so there was no need to 'switch off'. The value of E determined from the y-intercept is lower but only by about 5%. This value is less reliable than a direct measurement of E because of any scatter in the plotted points. However, the value obtained from the y-intercept should be used with the gradient to determine R to validate the theory.

Chapter 8: Waves

Chapter outline

This chapter relates to Chapter 13: Waves and Chapter 15: Stationary waves, in the coursebook.

In this chapter learners will complete investigations on:

- 8.1 Stationary waves on a wire carrying a current
- 8.2 Inverse-square law for waves from a point source
- 8.3 Refraction of light waves by a lens.

Practical Investigation 8.1: Stationary waves on a wire carrying a current

Skills focus

See the Skills grids at the front of this book for details of the skills developed and used in this investigation.

Duration

The practical work will take about 30 minutes; the analysis will take about 30 minutes.

Preparing for the investigation

- In this investigation the learner sets up a stationary wave in a wire. The wire is made to vibrate by positioning the wire in a magnetic field and passing an alternating current through it.
- The a.c. frequency is fixed at mains frequency, but the learner can vary the tension in the wire to change the wavelength of the stationary wave.
- The magnetic field is produced by two Magnadur magnets attached to a steel yoke, as shown in Figure 8.1.

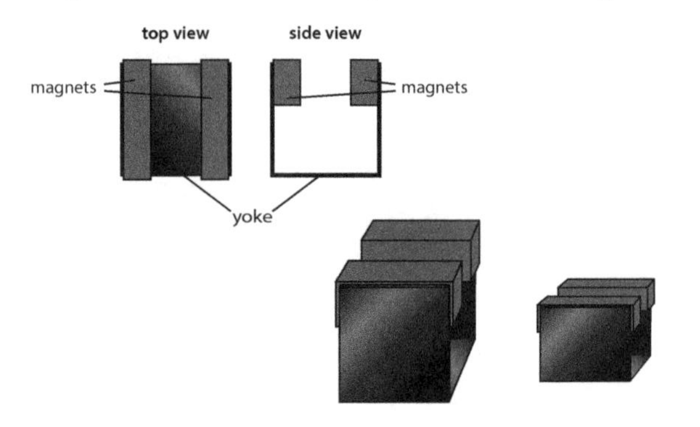

Figure 8.1

Equipment

Each learner or group will need:

- pulley wheel to clamp to edge of bench
- 1.2 m length of 30 swg constantan wire (copper wire of the same 0.315 mm diameter can be used as an alternative)
- triangular prism (glass or wood) tall enough to touch the wire at any position along the length
- two Magnadur magnets and a steel yoke (see Figure 8.1)
- 2 V mains frequency a.c. power supply. The mains frequency value should be written clearly on the top of the supply.
- two connecting leads, each with a clip at one end
- 100 g mass hanger
- two 100 g slotted masses and a 50 g slotted mass
- metre rule with a millimetre scale
- A4 sheet of dark-coloured paper
- See Figure 8.1 in the workbook for the arrangement. The apparatus should be partially assembled for the learners: the wire should be clamped between the wooden blocks and the mass hanger should be securely fixed to the other end of the wire.

Safety considerations

- As the wire is under tension from the start, learners must wear eye protection.
- There is no safety issue with the very low voltage electrical supply.

Carrying out the investigation

- A poor connection between a clip and the constantan wire can lead to weak vibrations which are harder to detect. Squeezing the jaws of the clip together usually cures this problem.

  In step **8** the learners may move the prism position too quickly when searching for the stationary wave. They need to understand that the amplitude takes time to build up at the resonant length.

  Some learners will need to be reminded of the need to include a quantity and correct unit for the column headings in all their tables.

  If some learners complete the task quickly they could be asked to write a description of a much simpler procedure for finding the value of μ for the wire.

Sample results

See Table 8.1 (shaded section)

M / kg	L / m	λ / m	λ^2 / m^2
0.100	**0.391**	0.782	0.612
0.150	**0.471**	0.942	0.887
0.200	**0.542**	1.084	1.175
0.250	**0.606**	1.212	1.469
0.300	**0.657**	1.314	1.727
0.350	**0.712**	1.424	2.028

Table 8.1

Answers to the workbook questions (using the sample results)

a, b See Table 8.1 (unshaded section).

c, d See Figure 8.2.

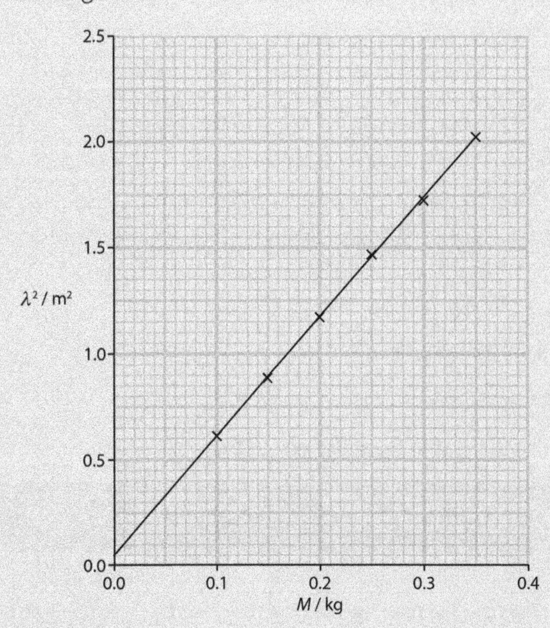

Figure 8.2

e Gradient = 5.65 and intercept = 0.044

f The player changes the wavelength by holding the string down at a different distance from its end.

g $\mu = \dfrac{g}{\text{gradient} \times f^2}$

$= \dfrac{9.81}{5.65 \times 50^2}$

$= 6.95 \times 10^{-4}\,\text{kg m}^{-1}$

Either measure wire diameter then calculate cross-sectional area A. Look up density ρ, then calculate μ using $\mu = A\rho$.

or Cut a 1.00 m length of the wire and find its mass μ using a top-pan balance with a precision of 0.01 g or better.

or Cut a 10.00 m length of the wire and find its mass using a top-pan balance with a precision of 0.1 g then divide by 10 to get μ.

Practical Investigation 8.2: Inverse-square law for waves from a point source

Skills focus

See the Skills grids at the front of this book for details of the skills developed and used in this investigation.

Duration

The practical work will take about 20 minutes and the analysis will take about 30 minutes.

Preparing for the investigation

- In this practical exercise the learner uses a light dependent resistor to monitor the light level (illuminance) at different distances from a small lamp. A graph is plotted to see if the data obeys an inverse-square law.
- The 6 V lamp is run at 9 V to provide a brighter point source.

Equipment

Each learner or group will need:

- A4 sheet of black paper rolled into a tube big enough for the rule to be pushed inside. It should be held in this shape using tape.
- 6 V 0.05 A miniature filament lamp (torch bulb) in a holder with connecting leads attached. It should be mounted inside the end of the tube as shown in Figure 8.3 in the workbook. Black tape should be used to seal it in position and exclude unwanted light.
- additional lamp (to be used when estimating the filament position inside the bulb)
- 9 V power supply
- two connecting leads
- light dependent resistor (LDR) type ORP12 with extended connecting leads. It should be attached to the zero end of a half-metre rule with a millimetre scale using tape as shown in Figure 8.3 in the workbook.
- ruler with a millimetre scale
- ohmmeter
- digital caliper.

Safety considerations

- The lamps have glass domes. If handled roughly these could break and cause cuts, so step **2** should be carried out carefully.

Carrying out the investigation

- In step **4** learners are left to choose their own positions of the LDR for taking readings. They should be advised to spread them so that a large part of the available range is used, but with larger changes as *B* increases (this avoids too much bunching of points at one end of the graph; see Sample results).

Learners may need to be reminded to record all their length readings to the nearest mm (even the *B* values which the learner can set to a whole number of cm).

Some learners will need to be reminded of the need to include a quantity and correct unit for the column headings in all their tables

If more able learners have finished the investigation and are familiar with log–log graphs, they could check the equation given in step **c** by deriving it themselves from the manufacturer's characteristic given in workbook Figure 8.6.

Common learner misconceptions

- It is good practice to keep the ohmmeter on the same range setting (20 kΩ) throughout the experiment, even at low resistances. Changing range sometimes causes jumps in the sequence of data.

Sample results

A = 26.5 cm　　　　*E* = 1.3 cm

Also see Table 8.2 (shaded sections).

B / cm	*x* / cm	$\frac{1}{x^2}$ / cm^{-2}	$\frac{1}{x^2}$ / m^{-2}	*R* / kΩ	*L* / lux
21.0	6.8	0.021 63	216.3	0.86	1687
19.0	8.8	0.012 91	129.1	1.22	1071
17.0	10.8	0.008 57	85.7	1.67	712
15.0	12.8	0.006 10	61.0	2.09	532
13.0	14.8	0.004 57	45.7	2.61	398
9.0	18.8	0.002 83	28.3	3.71	252
1.0	26.8	0.001 39	13.9	6.26	128

Table 8.2

Answers to the workbook questions (using the sample results)

a, b, c See Table 8.2 (unshaded section)

d, e See Figure 8.3.

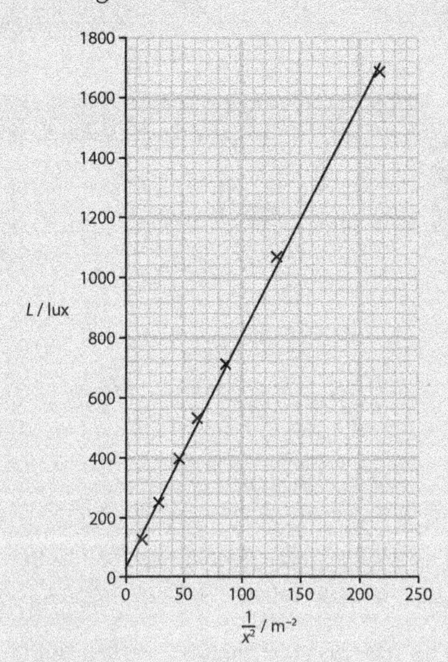

Figure 8.3

f Gradient = 7.71 and intercept = 44.2

g For proportionality points must lie close to a straight line which passes close to the origin.

Gradient of workbook Figure 8.6 $= \frac{-(6.25-3)}{(2-(-2))} = -0.8$

Intercept of workbook Figure 8.6 = 4.62

So equation is $\log R = -0.8 \log L + 4.625$

Giving $R = \frac{104.62}{L^{0.8}}$

or $L = \left[\frac{42k}{R}\right]^{1.3}$

or for L in lux $L = 10.746 \times \left[\frac{42k}{R}\right]^{1.3}$

Practical Investigation 8.3: Refraction of light waves by a lens

Skills focus

See the Skills grids at the front of this book for details of the skills developed and used in this investigation.

Duration

The practical work will take about 20 minutes; the analysis will take about 40 minutes.

Preparing the investigation

- The investigation does **not** involve a graph. Instead, it measures two sets of results and compares how well they fit a suggested relationship.
- In this exercise the learner measures the dimensions of a lens to find the radius of curvature of its surfaces and produces an image on a screen to finds its focal length. These values are used with the lens makers' formula to find the refractive index.
- An LED torch is used as the bright object as it gives a clear image on the screen even in a brightly lit room.

Equipment

Each learner or group will need:

- two 50 mm diameter biconvex lenses, one of focal length 10 cm and the other of focal length 20 cm
- digital calipers
- white screen of A4 size with a block of wood attached so that it can stand vertically on the bench
- small torch using at least three LEDs as its light source
- metre rule.

Safety considerations

There are no special safety issues with this experiment.

Carrying out the investigation

- When using the torch and screen in step **3** the learner must position them carefully at the stated distance apart. The axis of the lens must be kept in line with the torch and screen as it is moved to find the image.

  In step **2** the measurement of *E* is difficult as the jaws of the calipers tend to slide off the glass surfaces. Learners may find it easier to position the jaws close to the edge without making contact.

  If there is time, learners could measure *u* and *v* for different torch positions to confirm that the value of *f* remains the same.

Sample results

See Tables 8.3 and 8.4 (shaded sections).

	D / mm	*T* / mm	*E* / mm	*C* / mm	*R* / mm
Thinner lens	50.00	4.31	1.71	2.60	241
Thicker lens	50.00	7.86	2.41	5.45	116

Table 8.3

	u / cm	*v* / cm	*f* / cm	*R* / cm	η
Thinner lens	32.0	58.0	20.6	24.1	1.58
Thicker lens	11.2	78.8	9.81	11.6	1.59

Table 8.4

Answers to the workbook questions (using the sample results)

a See Table 8.3 (unshaded section).

b See Table 8.4 (unshaded section).

c Percentage difference between η values = 0.63%

d Using 0.5 cm as the absolute uncertainty for *u* gives percentage uncertainty in $u = 100 \times \frac{0.5}{11.2} = 4.5\%$

e The variation in η values could be due to variation in the data because the percentage uncertainty in the data is greater than the percentage difference in η.

f Example: *E* was difficult to measure as the callipers could only grip the curved surfaces rather than the edge.

g Example: The lens could be rolled along a surface (e.g. carbon paper on white paper, or modelling clay) to leave a mark, and then the width of the mark could be measured.

Chapter 9: Planning and data analysis

Chapter outline

This chapter relates to Chapter P2: Planning, analysis and evaluation in the coursebook.

In this chapter learners will complete investigations on:

- 9.1 Planning data analysis
- 9.2 Treatment of uncertainties
- 9.3 Planning investigation into how the acceleration of a vehicle rolling down an inclined plane varies with the angle of the plane
- 9.4 The acceleration of a vehicle rolling down an inclined plane
- 9.5 Planning investigation into how the current in an LDR varies with the distance from a light source
- 9.6 The resistance of an LDR
- 9.7 Planning investigation into how the electromotive force (emf) of a photovoltaic cell varies with the thickness of an absorber

Practical Investigation 9.1: Planning data analysis

Skills focus

See the Skills grids at the front of this book for details of the skills developed and used in this investigation.

Duration

The exercise will take 30 minutes for learners who have been prepared in the use of graphical analysis and the use of logarithms.

Preparing for the exercise

Learners need to:

- understand the terms independent variable and dependent variable
- identify quantities that need to be kept constant for a fair test
- understand the equation for a straight line relates to quantities plotted on a graph
- be able to use logarithms to base 10.

Carrying out the investigation

Encourage learners to write down the steps in any rearrangement of the expression given. This assists in the identification of variables as well as identifying quantities that the gradient and y-intercept represent.

If learners are having difficulty with rearranging equations, further work could be set as a homework exercise.

Projectile motion: Learners could consider the advantages and disadvantages of plotting other graphs such as R^2 against $4L \sin \theta$.

Learners could consider why the graph of R^2 against $4Lh \sin \theta$ should not be plotted.

Current in a filament lamp: learners could consider the difference in the circuits needed to vary the potential difference across the lamp compared to the current through the lamp.

- Learners could determine the constants p and q for a graph of $\lg V$ on the y-axis and $\lg I$ on the x-axis.

Common learner misconceptions

- It may be necessary to explain how to rearrange an expression into the equation of a straight line.
- Learners sometimes struggle with square roots.

Part 1: Projectile motion

Variables

Learners should identify:

- dependent variable: horizontal distance R from the end of the table to the point P
- independent variable: distance L along the slope from where the ball is released
- quantities to be kept constant: the height h of the table and the angle θ of the slope to the bench.

Data analysis

Suitable answers could be as follows:

- graph to plot: R^2 on the y-axis and L on the x-axis
- equation of straight line: $R^2 = (4h\sin\theta)L$
- gradient = $4h\sin\theta$
- $h = \frac{gradient}{4\sin\theta}$

Alternative answers could be:

- Graph to plot: R on the y-axis and $\sqrt{L}$ on the x-axis.
- Equation of straight line: $R = \sqrt{4h\sin\theta}\sqrt{L}$
- Gradient = $\sqrt{4h\sin\theta}$
- $h = \frac{\text{gradient}^2}{4\sin\theta}$

Part 2: Current in a filament lamp

Variables

Learners should identify:

- dependent variable: current I
- independent variable: potential difference V

Data analysis

Suitable answers could be as follows:

- graph to plot: $\lg I$ on the y-axis and $\lg V$ on the x-axis.
- equation of straight line: $\lg I = q \lg V + \lg p$
- gradient: q $\quad$ y-intercept: $\lg P$
- $p = 10^{y\text{-intercept}}$ $\quad$ q = gradient

Practical Investigation 9.2: Treatment of uncertainties

Skills focus

See the Skills grids at the front of this book for details of the skills developed and used in this investigation.

Duration

Each exercise will take 30 minutes for learners who have been prepared in the treatment of uncertainties.

Preparing for the investigation

Learners need to understand:

- the terms absolute uncertainty and percentage uncertainty
- the rules for adding absolute uncertainties and percentage uncertainties.

Learners could carry out these exercises practically.

Carrying out the investigation

Encourage learners to write down the steps taken.

Discussion of maximum and minimum methods helps learners to understand the rules for adding percentage uncertainties and absolute uncertainties.

If learners are having difficulty with the treatment of uncertainties, further exercises could be set for homework.

Learners could determine the maximum and minimum density of the metal ball and thus show the absolute uncertainty is similar.

Common learner misconceptions

- Learners are often confused about when to add absolute uncertainties and when to add percentage uncertainties.
- When finding the maximum or minimum value of a quotient, learners will often incorrectly apply the maximum and minimum values.
- Learners do not realise that when a quantity is given to a power, it is in effect multiplying the quantity and so percentage uncertainties should be **added**.

Answers to the workbook questions

Part 1: Density of a liquid

a Mass = 279 ± 2 g

b Max. mass = max mass of beaker and cooking oil – min. mass of empty beaker

Max. mass = 604 g – 323 g = 281 g

Min. mass = min. mass of beaker and cooking oil – max. mass of empty beaker

Min. mass = 602 g – 325 g = 277 g

c Absolute uncertainty = half the range
$= 0.5 \times (281 - 277)\ \text{g} = \pm 2\ \text{g}$

d $\text{Density} = \dfrac{\text{mass}}{\text{volume}} = \dfrac{279}{300} = 0.93\ \text{g cm}^{-3}$

e $\text{Percentage uncertainty in mass} = \dfrac{2}{279} \times 100 = 0.7\%$

f $\text{Percentage uncertainty in volume} = \dfrac{5}{300} \times 100 = 1.7\%$

g Percentage uncertainty in density = 0.7% + 1.7% = 2.4%

h Absolute uncertainty = $0.024 \times 0.93 = \pm 0.02\ \text{g cm}^{-3}$

i $\text{Max. }density = \dfrac{\text{max. mass}}{\text{min. volume}} = \dfrac{281}{295} = 0.95\ \text{g cm}^{-3}$

j $\text{Min. }density = \dfrac{\text{min. mass}}{\text{max. volume}} = \dfrac{277}{305} = 0.91\ \text{g cm}^{-3}$

k Absolute uncertainty = $0.5 \times (0.95 - 0.91) = \pm 0.02\ \text{g cm}^{-3}$

Part 2: Density of metal sphere

a Mean diameter = 1.6125 cm = 1.61 cm

Note: measured diameters are given to 3 significant figures, so the calculated diameter should also be given to 3 significant figures.

Absolute uncertainty = half the range
$= 0.5 \times (1.63 - 1.59)\ \text{cm} = \pm 0.02\ \text{cm}$

Diameter = $1.61 \pm 0.02\ \text{cm}$

b $V = \dfrac{\pi \times 1.61^3}{6} = 2.19\ \text{cm}^3$

c $\text{Percentage uncertainty in } d = \dfrac{0.02}{1.61} \times 100 = 1.24\%$

Percentage uncertainty in $V = 3 \times 1.24\% = 3.7\%$ or 4%

d Absolute uncertainty = half the range
$= 0.5 \times (281 - 277)\ \text{g} = \pm 2\ \text{g}$

e $\text{Density} = \dfrac{\text{mass}}{\text{volume}} = \dfrac{19}{2.19} = 8.7\ \text{g cm}^{-3}$

Note: Two significant figures since mass is given to 2 s.f. and diameter to 3 s.f.

f $\text{Percentage uncertainty in mass} = \dfrac{1}{19} \times 100 = 5.3\%$

Percentage uncertainty in density = 5.3% + 3.7% = 9%

g Absolute uncertainty = $0.09 \times 8.7 = \pm 0.8\ \text{g cm}^{-3}$

h Density = $8.7 \pm 0.8\ \text{g cm}^{-3}$

Practical Investigation 9.3: Planning

Investigation into how the acceleration of a vehicle rolling down an inclined plane varies with the angle of the plane

Skills focus

See the Skills grids at the front of this book for details of the skills developed and used in this investigation.

Duration

The planning investigation will take about 40 minutes.

Preparing for the investigation

- It may be helpful for teachers to show an inclined plane and a toy car or trolley rolling down the plane.

Learners need to:

- know about different types of variables and the control of quantities
- be able to draw graphs and calculate their gradient.

Variables

Learners should identify:

- the dependent variable is a, the acceleration of the toy car
- the independent variable is θ
- the variables to be controlled are using the same toy car and the same inclined plane (so that friction is constant).

Equipment

Each learner or group will need:

- trolley or toy vehicle
- stopwatch or electronic timer
- board/ramp for the inclined plane
- retort stand and clamps (or books) to support the inclined plane
- metre rule(s)

- protractor use a cardboard box or cushion for the vehicle to roll into at the bottom of the inclined plane
- light gate(s) to assist in the determination of a.

Method

A suitable apparatus set-up (see Figure 9.2 in the workbook). The method chosen by a learner could be as follows.

- Labelled diagram showing the slope supported.
- The diagram could also show how the angle could be measured with a protractor or the lengths measured to determine the angle by trigonometry. The diagram should also indicate the measurements that might be taken to determine the acceleration.

There are several methods to determine the acceleration of the toy car.

- Use a stopwatch or two light gates connected to a timer to time (t) the toy car for a set distance s:

$$a = \frac{2s}{t^2}$$

- Use a piece of card (with a gap in the top) as shown in Figure 9.1 and one light gate connected to a data logger to determine the acceleration directly. Appropriate length measurements should be indicated.

Figure 9.1

Extra detail, for example:

- detail on using light gate(s) connected to a timer
- method to ensure that the toy car rolls straight down the plane and does not veer to the side
- use slow motion, freeze-frame video, to find the extreme edge of motion
- expression to determine sin θ from lengths indicated on the diagram
- clean the slope
- use of card to block light gate.

Carrying out the investigation

- Learners may need help in realising how they can measure the value of the angle θ.
- Learners find the determination of acceleration practically difficult. The methods outlined above could be discussed.

Some learners will need help in starting the planning. A good approach may be to list all the necessary quantities to be measured, how each can be measured and what apparatus is needed. Learners should make sure each quantity is described and as much detail as possible is given. They should review their work at the end to make sure that if, for example, they have determined the angle by a trigonometric method, they have measured appropriate lengths and given the relationships.

Some learners, who have finished the investigation, can use a toy car and trial their method.

Learners could consider other methods of determining acceleration using light gates.

Common learner misconceptions

- Learners will often consider that the velocity at the end of the plan is given by the distance travelled divided by the time (average velocity) rather than final velocity.
- Learners may not understand that a light gate is a detector and needs to be connected to a timer or data logger. Furthermore, learners may not understand that to determine the acceleration, the length that is interrupting the beam needs to be measured. If a datalogger is used, then measurements will need to be entered into the datalogger.

Answers to the workbook questions

a Plot a graph of a against sin θ.

- Relationship is valid if the result is a straight line through the origin
- g = gradient

b The resultant force causing the acceleration is equal to the weight of vehicle (acting downwards) and is acting parallel to the plane.

weight parallel to the plane = $mg\sin\theta$

$F = ma = mg\sin\theta$

$\therefore a = g\sin\theta$

Practical Investigation 9.4: Investigation into the acceleration of a vehicle rolling down an inclined plane

Skills focus

See the Skills grids at the front of this book for details of the skills developed and used in this investigation.

Duration

The practical will take about 60 minutes.

Preparing for the investigation

- See details provided in Practical investigation 9.3.

Equipment

Each learner or group will need:

- trolley or toy car
- stopwatch
- inclined plane
- variable support for inclined plane
- metre rule
- protractor
- cardboard box
- light gate(s) to assist in the determination of a.

Carrying out the investigation

- Learners may need help in calculating uncertainties and the drawing of the worst acceptable line.
- Learners could also work out the percentage uncertainty using maximum/minimum methods.

Some learners will need help in determining the uncertainties.

Some learners will need to be reminded of the need to include a quantity and correct unit for the column headings in all their tables.

Learners who have finished the investigation could estimate the uncertainty in L and D and then determine the percentage uncertainty in their value of g.

Common learner misconceptions

- In the table of results, learners will need to take into account that the determination of the uncertainty from two values is found from **half** the range, while $\frac{1}{t^2}$ requires either a maximum minimum method or adding the percentage uncertainties and then finding the absolute uncertainty.
- Some learners will need to be reminded of the need to include a quantity and correct unit for $\frac{1}{t^2}$.

Sample results

$L = 180\,\text{cm}$ $D = 150\,\text{cm}$

	t / s			
h / cm	1st value	2nd value	average	$\frac{1}{t^2}$ / s^{-2}
25.0	1.55	1.49	1.52 ± 0.03	0.433 ± 0.017
30.0	1.35	1.41	1.38 ± 0.03	0.525 ± 0.023
35.0	1.25	1.31	1.28 ± 0.03	0.610 ± 0.029
40.0	1.23	1.17	1.20 ± 0.03	0.694 ± 0.035
45.0	1.10	1.16	1.13 ± 0.03	0.783 ± 0.042
50.0	1.04	1.10	1.07 ± 0.03	0.873 ± 0.049

Table 9.1

Answers to the workbook questions (using the sample results)

a, b See Table 9.1.

Note: All the height measurements have been recorded to the nearest millimetre and the raw time measurements have all been recorded to the nearest 0.01 second. In the average column, since the raw times are to 3 significant figures the average is also given to 3 significant figures and $\frac{1}{t^2}$ is given to 3 significant figures.

c, f See Figure 9.2.

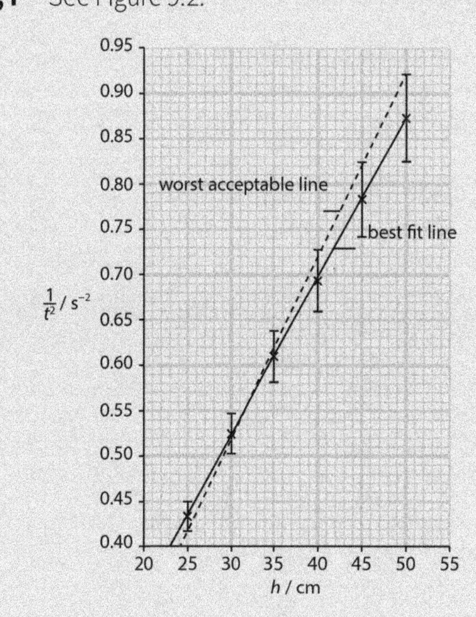

Figure 9.2

d $\frac{1}{t^2} = \frac{g}{2LD}h$

e Gradient $= \frac{g}{2LD}$

g Gradient = 0.0175 ($cm^{-1}s^{-2}$); gradient of worst line = 0.016 or 0.02 ($cm^{-1}s^{-2}$) uncertainty in gradient = ± 0.0025 ($cm^{-1}s^{-2}$)

h 945 cm s^{-2} or 9.45 m s^{-2}

i 14%

If there is no uncertainty in L and D, then the percentage uncertainty in g is the same as the percentage uncertainty in the gradient.

j There could be a systematic error in determining t.

There is also likely to be friction between the trolley and the inclined plane

k Ensure that the inclined plane is smooth and that the friction between the trolley and the inclined plane is minimised: oil the wheels.

Practical Investigation 9.5: Planning

Investigation into how the current in an LDR varies with the distance from a light source

Skills focus

See the Skills grids at the front of this book for details of the skills developed and used in this investigation.

Duration

The planning investigation will take about 40 minutes.

Preparing for the investigation

- It may be helpful for teachers to show a light dependent resistor.

Learners need to:

- know about different types of variables and the control of quantities
- be able to draw graphs and calculate their gradient.

Variables

Learners should identify:

- the dependent variable is I, the current in the LDR
- the independent variable is D, the distance from the light source
- the variables to be controlled are keeping the same intensity of the light source and the potential difference across the LDR.

Equipment

Each learner or group will need:

- lamp and power supply
- LDR
- power supply and ammeter connected to the LDR
- cardboard to shield external light
- learners might also suggest the use of an ammeter and variable resistor to ensure that the intensity of the light source is constant.

Method

A suggested method is as follows:

1. Labelled diagram showing two circuits (with correct circuit symbols); LDR connected to an ammeter and power supply.
2. Light source connected to a power supply.
3. The diagram could also show shielding of external light sources.

Extra detail, for example:

- a safety precaution linked to the bright light; not looking at it or using dark (sun)glasses or not touching the light source because it is hot
- further detail on measuring the distance D
- a method to ensure that the current in the light circuit is constant
- the given equation could be given in its logarithmic form: $\lg I = q \lg D + \lg p$.

Carrying out the investigation

- Learners may need help determining the logarithmic relationship.

Some learners will need help choosing an appropriate graph to plot.

Learners who have finished the investigation could trial their plan.

Common learner misconceptions

- Learners will often draw one circuit for both the light source and the LDR.

Answers to the workbook questions

a Plot a graph of $\lg I$ against $\lg D$.

- Relationship is valid if the result is a straight line with gradient q.
- q = gradient
- $p = 10^{y\text{-intercept}}$

Practical Investigation 9.6: Data analysis

Investigation into the resistance of an LDR

Skills focus

See the Skills grids at the front of this book for details of the skills developed and used in this investigation.

Duration

The data analysis and evaluation questions will take about 60 minutes.

Preparing for the investigation

- See details provided in Practical investigation 9.5.

Carrying out the investigation

- Learners may need help in calculating uncertainties and the drawing of the worst acceptable line.

  Some learners will need help in determining the uncertainties in both R and $\lg R$. Learners may also need help in the mathematical analysis.

  Learners who have finished the investigation could check other methods for determining the uncertainties in R.

Common learner misconceptions

- In the table of results, learners will need to obtain the uncertainty in R using a maximum minimum method or adding the percentage uncertainties and then finding the absolute uncertainty.

Answers to the workbook questions

a, b The voltmeter reading $V = 5.6 \pm 0.1$ V

D / cm	I / mA	lg (D / cm)	R / Ω	lg (R / Ω)
22.5	9.6 ± 0.2	1.352	583 ± 20	2.763 ± 0.017
34.5	5.2 ± 0.2	1.538	1080 ± 60	3.033 ± 0.024
51.0	3.0 ± 0.2	1.708	1870 ± 160	3.271 ± 0.037
64.0	2.2 ± 0.2	1.806	2550 ± 280	3.407 ± 0.047
81.5	1.6 ± 0.2	1.911	3500 ± 510	3.544 ± 0.062
96.5	1.3 ± 0.2	1.985	4300 ± 660	3.633 ± 0.075

Note: All the distance measurements (D) have been recorded to the nearest millimetre (3 significant figures) so the number of decimal places in lg (D/cm) is given to 3 decimal places.

$\lg R$ could be given to 2 decimal places since the data for both current and p.d. is given to 2 significant figures.

c, f See Figure 9.3.

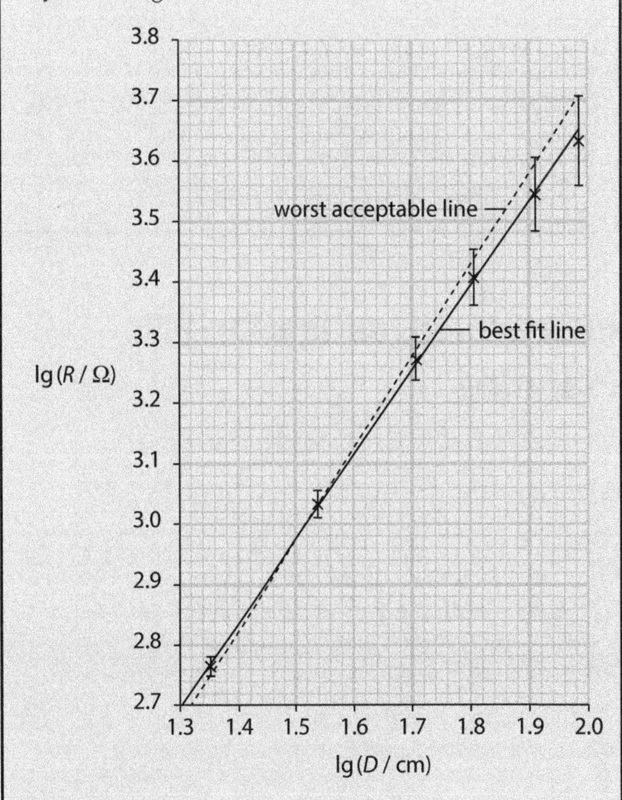

Figure 9.3

d $\lg R = J \lg D + \lg k$

e Gradient $= J$
y-intercept $= \lg k$

g Gradient = 1.4
Gradient of worst-fit line = 1.5 or 1.3
Uncertainty in gradient = ±0.1

h y-intercept = 0.87 y-intercept of worst-fit line = 0.69 or 1.05 Uncertainty in y-intercept = ±0.2

i $J = 1.4$ $k = 7.4$

j Uncertainty in J = ±0.1, uncertainty in k = ±2.5

k D is measured from the surface of the lamp to the surface of the LDR so D measured is smaller than it should be.

l Measure the distance between the filament of the lamp and the surface of the lamp and estimate the distance to the LDR.

m If D is measured too short, this means that the D values should all be larger so $\log D$ should be larger.

The graph is likely to move to the right which means the gradient may change and the y-intercept may be smaller. Since for fixed increase in D the effect is greater for lower values of D, the gradient is likely to be steeper.

J is likely to be larger and k will be smaller.

Practical Investigation 9.7: Planning

Investigation into how the electromotive force (emf) of a photovoltaic cell varies with the thickness of an absorber

Skills focus

See the Skills grids at the front of this book for details of the skills developed and used in this investigation.

Duration

The planning investigation will take about 40 minutes.

Preparing for the investigation

- It may be helpful for teachers to show a photovoltaic cell.

Learners need to:

- know about different types of variables and the control of quantities
- be able to draw graphs and calculate their gradient.

Variables

Learners should identify:

- The dependent variable is E the generated e.m.f. from the photovoltaic cell.
- The independent variable is n, the number of plastic slides.
- The variables to be controlled are keeping the thickness of slide and the intensity of the light source constant. The distance from the light source to the photovoltaic cell should also be kept constant.

Equipment

A suggested equipment list is:

- lamp and power supply
- power supply and ammeter connected to the LDR
- photovoltaic cell
- retort stand to support lamp above slides
- micrometer or Vernier callipers to measure thickness of the slide
- cardboard to shield external light
- learners might also suggest the use of an ammeter and variable resistor to ensure that the intensity of the light source is constant.

Method

A suggested method is as follows.

Labelled diagram showing two circuits (with correct circuit symbols):

- photovoltaic cell connected to a voltmeter
- light source connected to a power supply
- diagram could also show shielding of external light sources
- thickness of a slide measured using an appropriate instrument.

Extra detail, for example:

- detail on measuring thickness t e.g. repeat readings and average
- a safety precaution linked to the bright light: not looking at it or using dark (sun)glasses or not touching the light source because it is hot
- a method to the light source is fixed
- plastic slides are cleaned before use
- the equation could be given in its logarithmic form, i.e. $\ln E = -ktn + \ln E_0$.

Carrying out the investigation

- Learners may need help determining the logarithmic relationship.

  Some learners will need help choosing an appropriate graph to plot.

  Learners who have finished the investigation could trial their plan.

Common learner misconceptions

- Learners will often draw one circuit for both the light source and the photovoltaic cell.

Answers to the workbook questions

a Plot a graph of $\ln E$ against n.

Relationship is valid if the result is a straight line:

$$k = -\frac{\text{gradient}}{t}$$

$$E_0 = e^{y\text{-intercept}}$$

Chapter 10: Circular motion and gravitational fields

Chapter outline

This chapter relates to Chapter 17: Circular motion and Chapter 18: Gravitational fields, in the coursebook.

In this chapter learners will complete investigations on:

- 10.1 Circular motion
- 10.2 Planning investigation into the conical pendulum
- 10.3 Data analysis investigation into the conical pendulum
- 10.4 Data analysis investigation into planetary motion
- 10.5 Data analysis investigation into gravitational potential

Practical Investigation 10.1: Circular motion

Skills focus

See the Skills grids at the front of this book for details of the skills developed and used in this investigation.

Duration

The practical work will take 30 minutes; the analysis and evaluation questions will take 30 minutes.

Preparing for the investigation

Learners need to:

- know the equations for centripetal force in terms of angular velocity and period
- be able to calculate uncertainty from several readings of the same quantity
- be able to draw graphs and calculate their gradient
- be able to time the circular motion as a rubber bung rotates in a horizontal circle.

Equipment

Each learner or group will need:

- short length of tube, glass or plastic about 10 cm long, such as the casing of a pen. If glass is used the ends should be burred over to avoid scratching the skin.
- stopwatch
- 100 g masses or 50 g masses or washers. If washers are used they should be identical and their mass determined with the use of a balance.
- rubber bung with hole. The mass of the rubber bung should be about 30 g.
- string (about 1 m). The string should be firmly tied to the rubber bung passing through the hole in the bung so that the rubber bung cannot break away. The string then passes through the short length of tube. Tie one of the masses with a mass slightly greater than the mass of the bung to the lower end of the string.
- metre rule
- marker pen to mark the string.

Safety considerations

- Eye protection must be worn by all during the experiment in case the string holding the rubber bung breaks.
- Ample space for each experiment must be available so the bung can be rotated without any risk to other people or apparatus.

Carrying out the investigation

- Some learners find it difficult to rotate the glass tube and keep the mass rotating in an approximately horizontal circle. It is best to start with a slow rotation and increase speed until the required radius is achieved. With a hanging weight whose mass is about 3 times that of the rubber bung and a radius of about 70 cm, the period is about 1 s.
- The weights must hang freely and not rise to touch the bottom of the tube because then the centripetal force

will not be equal to the weight of the hanging masses. You may need to demonstrate this procedure first.

- A second learner measures the time for 10 oscillations. You may need to show learners how this is done, counting zero and starting the stopwatch as the mass passes one point and then 1, 2 etc. and stopping the stopwatch on the count of 10.

The calculation of uncertainty in T should be familiar to most learners but calculation of the uncertainty in T^{-2} may need to be explained: the percentage uncertainty in T^{-2} is twice the percentage uncertainty in T. Alternatively, calculations of the maximum and minimum values of T^{-2} can be used to find the uncertainty.

If learners are struggling with uncertainties then they can perform the whole experiment without dealing with uncertainty and draw only the best-fit line. Further analysis on uncertainty can then be done together as a group or class, or even as a homework exercise.

It is also possible to change the value of R with a constant weight on .the string and repeat the measurements. They could then consider the relationship between T^{-2} and R for a constant value of m.

Common learner misconceptions

- It may be necessary to explain how the weight of the hanging masses provides the centripetal force, as it is not immediately obvious.

Sample results

Table 10.1 provides results that learners could obtain in the investigation. Only the values for R, T_{10} and m should be given.

$R = 0.76$ m

	Time for 10 oscillations T_{10} / s				
m / kg	1st value	2nd value	average	T / s	T^{-2} / s^{-2}
0.10	10.17	8.67	9.42 ± 0.8	0.942 ± 0.08	1.13 ± 0.2
0.20	7.82	6.30	7.06 ± 0.8	0.706 ± 0.08	2.01 ± 0.5
0.30	5.33	5.80	5.57 ± 0.2	0.557 ± 0.02	3.23 ± 0.2
0.40	5.28	4.78	5.03 ± 0.3	0.503 ± 0.03	3.95 ± 0.4
0.50	4.15	4.47	4.31 ± 0.2	0.431 ± 0.02	5.38 ± 0.4

Table 10.1

Answers to the workbook questions (using the sample results)

a, b See Table 10.1.

c, f See Figure 10.1

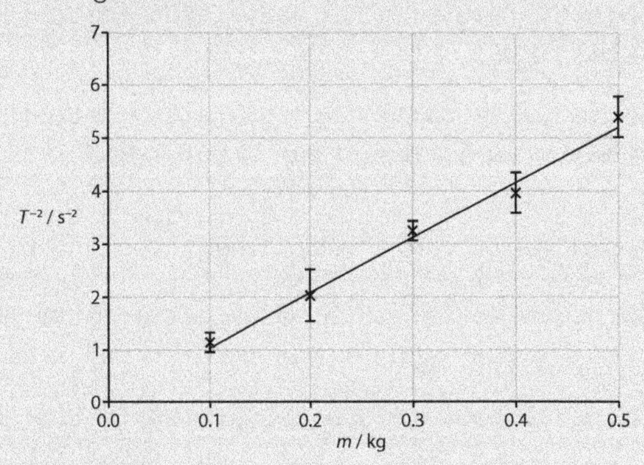

Figure 10.1

d $T^{-2} = \dfrac{mg}{4\pi^2 MR}$

e Gradient $= \dfrac{g}{4\pi^2 MR}$

g Approximate gradients using sample results:

Gradient of best-fit line = 10.5

Gradient of worst-fit line = 10.8

Uncertainty in gradient = 0.3 *or* 3%

h $M = \dfrac{g}{\text{gradient} \times 4\pi^2 R}$; = 0.031 kg; uncertainty 3%

i The force in the string is tension. Friction means that the weight is larger than the force of tension acting on the rubber bung and the value of T^{-2} is smaller.

j The string is pulling on the mass with a force. The mass is pulling on the string with an equal and opposite force. The hanging weights are pulled to the Earth by the force of gravity. The weights pull upwards on the Earth with an equal and opposite force.

Practical Investigation 10.2: Planning

Investigation into the conical pendulum

Skills focus

See the Skills grids at the front of this book for details of the skills developed and used in this investigation.

Duration

This planning investigation will take about 40 minutes.

Preparing for the investigation

It may be helpful to show a conical pendulum and to explain how the forces acting on a conical pendulum produce the formula $\cos\theta = \frac{g}{L\omega^2}$.

Learners will design a laboratory experiment to test the relationship between θ and ω.

Learners need to:

- know the formula for circular motion: $T = \frac{2\pi}{\omega}$
- know about different types of variables and the control of quantities
- be able to draw graphs and calculate their gradient
- be able to combine uncertainties.

Variables

Learners should identify:

- the independent variable is ω, the angular velocity of the aeroplane or the speed of the aeroplane
- the dependent variable is θ
- the variables to be controlled are the length of the string and the mass of the aeroplane.

Equipment

A suggested equipment list is:

- stopwatch or electronic timer
- camera and video recorder, e.g. mobile telephone
- toy aeroplane
- means of viewing video, e.g. computer, projection screen
- ruler for measuring length of string or lengths on screen to calculate θ or a protractor
- a protective screen in case the aeroplane detaches from the string.

Method

A suggested method includes:

1. Labelled diagram showing camera photographing the motion of the toy aeroplane with the stopwatch also in vision so that the time for each frame can be found. Alternatively, a light shone at the plane so that its shadow can be seen on a screen.
2. Vary the speed of the plane or its angular velocity ω (e.g by altering propeller motor) and measure θ.
3. Measure T (e.g. stopwatch to time a number of rotations, light gates connected to a timer / stopwatch shown in video) or the frequency f with light gates and computer.
4. Use a fiducial mark in timing or use light gates perpendicular to the motion of the plane.
5. Measure angle θ with the use of a protractor or a ruler for measurement using trigonometry.

Extra detail, for example:

- adjust motor speed to produce measurable changes in θ
- video projected onto a screen to measure θ
- use slow motion, freeze-frame video, to find the extreme edge of motion
- $\cos\theta$ calculated from $\frac{h}{l}$ shown on diagram
- time at least 10 rotations.

Results

Learners plan to record observations such as shown in Table 10.2 in Practical investigation 10.3.

Carrying out the investigation

- Learners may need help in realising how they can measure the value of the angle θ and they may have no experience of using a video which is played back a frame at a time. Alternatively, the shadow of a conical pendulum can be cast on a screen by a point source of light and learners can be invited to use a protractor to measure the angle on the screen.

  Some learners will need help in starting the planning. A good approach may be to list all the necessary quantities to be measured, how each can be measured and what apparatus is needed. Learners should make sure each quantity is described in as much detail as possible. They should review their work at the end to make sure that if, for example, they have mentioned the length of the string in the analysis, they have measured it and included the necessary apparatus.

Some learners, who have finished the investigation, can use a mass on a string as a conical pendulum to take some measurements for themselves to see some of the problems involved.

Common learner misconceptions

- Learners may not understand that the time required is for a complete rotation and not just for the movement of the plane from one side to the other.

Answers to the workbook questions

a Learners plan to analyse data, e.g. ω calculated from $\omega = 2\pi f = \frac{2\pi}{T}$.

Plot a graph of $\cos\theta$ against $\frac{1}{\omega^2}$.

The relationship is valid if the result is a straight line through the origin with gradient $\frac{g}{L}$.

b $T\sin\theta = mr\omega^2$

Thus $T\sin\theta = mr\omega^2$ and $\frac{Tr}{L} = mr\omega^2$ and $T = mL\omega^2$

Vertically $T\cos\theta = mg$ so $T = \frac{mg}{\cos\theta} = mL\omega^2$ and $\cos\theta = \frac{g}{L\omega^2}$.

Practical Investigation 10.3: Data analysis

Investigation into the conical pendulum

Skills focus

See the Skills grids at the front of this book for details of the skills developed and used in this investigation.

Duration

The data analysis and evaluation questions will take about 60 minutes.

Preparing for the investigation

See details provided in Practical investigation 10.2.

Learners will use the sample results to test the relationship between θ and ω.

Carrying out the investigation

- Learners may need help in calculating uncertainties, particularly with angles involved.
- Teachers might suggest that the uncertainty in only one value of $\cos\theta$ is calculated and that this is used for all the other values of $\cos\theta$.

Common learner misconceptions

Learners may not understand how to use an uncertainty in an angle to find the uncertainty in the cosine of that angle and may think that the cosine has the same percentage uncertainty as the angle. Teachers may need to show learners that they should calculate two values of the cosine to determine the uncertainty.

Answers to the workbook questions

a Gradient $= \frac{g}{L}$

b

θ / °	$\cos\theta$	Time for 10 circuits T_{10} / s	T / s	ω / s^{-1}	$\frac{1}{\omega^2}$ / s^2
10 ± 1	0.985 ± 0.007	14.1	1.41	4.46	0.0504
22 ± 1	0.927 ± 0.014	13.7	1.37	4.59	0.0475
32 ± 1	0.848 ± 0.019	13.1	1.31	4.80	0.0435
42 ± 1	0.743 ± 0.024	12.3	1.23	5.11	0.0383
53 ± 1	0.602 ± 0.028	11.0	1.10	5.71	0.0306
71 ± 1	0.326 ± 0.033	8.1	0.81	7.76	0.0166

Table 10.2

c, d See Figure 10.2

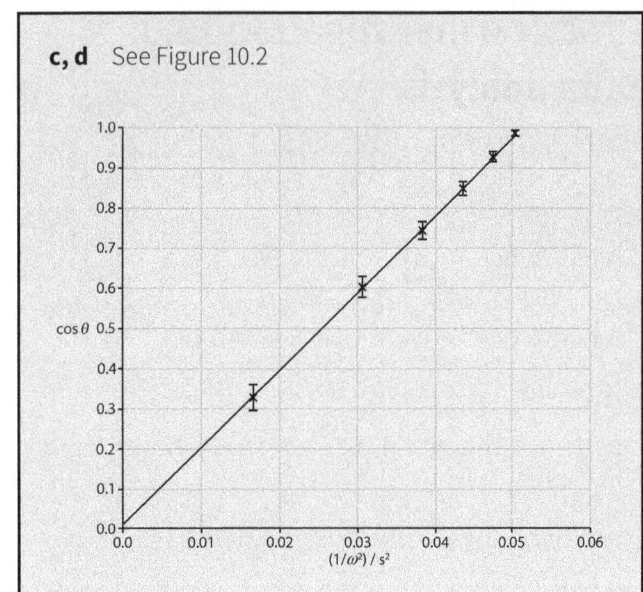

Figure 10.2

e Gradient = 19.2 s^{-2}; uncertainty 0.2 – 0.4 s^{-2}

f $L = \frac{g}{\text{gradient}} = \frac{9.81}{19.2} = 0.511\,\text{m}$

Uncertainty has same percentage as gradient,

e.g. $0.2 \times \frac{100}{19.2} = 1\text{–}2\%$

g $\cos\theta = 1$ when $\theta = 0$ and $\omega^{-2} = 0.52$, $\omega = 1.39\,s^{-1}$.

$T = \frac{2\pi}{\omega} = 4.53\,\text{s}$

Practical Investigation 10.4: Data analysis

Investigation into planetary motion

Skills focus

See the Skills grids at the front of this book for details of the skills developed and used in this investigation.

Duration

The analysis and evaluation questions will take about 45 minutes.

Preparing for the investigation

- It may be helpful if learners are shown how Kepler's third law is derived from Newton's law of gravitation.
- Learners will use data to investigate the relationship between the period T of the orbit of a planet and its distance R from the Sun.

Learners need to be able to:

- use logarithms to base 10
- use the relationships $\lg(a^x) = x\lg a$ and $\lg(a \times b) = \lg a + \lg b$
- draw graphs and find gradients including worst-fit lines.

Carrying out the investigation

If more able learners have finished the investigation, suggest they help other groups who may be struggling.

Common learner misconceptions

- Learners may write down a logarithm to the same number of significant figures as are present in the number. The number of decimal places for the logarithm of a number to base 10 should be equal to the number of significant figures of the number. This is because the power of 10 in the number becomes the part of the logarithm before the decimal place.

  In part **h**, learners are asked to find a value for k, where k is the intercept on the graph. However, since the x-axis is not at $y = 0$ the intercept cannot be read from the graph and must be calculated using one point and the calculated gradient.

Answers to the workbook questions

a

Planet	R / 10^{10} m	T / 10^6 s	lg (R / m)	Maximum value of lg (R / m)	lg (T / s)
Mercury	5.8 ± 0.4	7.59	10.763	10.792	6.880
Venus	10.8 ± 0.1	19.4	11.033	11.037	7.288
Earth	15.0 ± 0.0	31.5	11.176	11.176	7.498
Mars	22.8 ± 0.1	59.3	11.358	11.358	7.773
Jupiter	78 ± 3	374	11.892	11.908	8.573
Saturn	140 ± 40	929	12.146	12.255	8.968

Table 10.3

b, c See Figure10.3

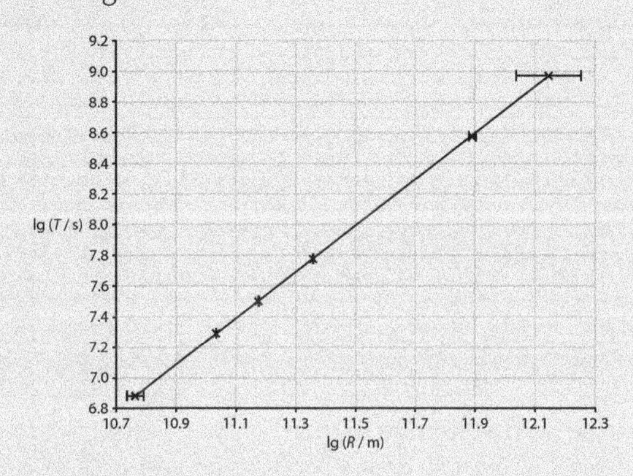

Figure 10.3

d Gradient of line of best fit = 1.50; gradient of line of worst fit = 1.46 approximately

e Uncertainty in gradient = 0.04 using values in part **d**.

f $\lg T = n \lg R + \lg k$

g $n = 1.50 \pm 0.04$

h $\lg k = 7.498 - 1.50 \times 11.176 = -9.266$

$k = 5.42 \times 10^{-10}$ (Note: a slightly different value of n changes the value of k significantly.)

i $2 \lg T = 3 \times \lg R + \lg \left[\dfrac{4\pi^2}{GM_s}\right]$

j The value for n should be 1.50 and the value obtained was within the uncertainty of this value.

k 2.0×10^{30} kg (Note: a slightly different value of n changes the value of M_S significantly.)

Practical Investigation 10.5: Data analysis

Investigation into gravitational potential

Skills focus

See the Skills grids at the front of this book for details of the skills developed and used in this investigation.

Duration

The analysis and evaluation questions will take 45 minutes.

Preparing for the investigation

- Learners will use data to check the formula for gravitational potential and to calculate the mass of the Earth.
- It may be helpful in understanding the theory for the ideas of gravitational potential and total energy to be covered in class. Learners may also research the Apollo flights and the telemetry data sent back to Earth form the spacecraft.

Carrying out the investigation

The first and last row of the results table have been calculated already. Teachers may need to check learners' attempts before they proceed to plot the graph.

More able learners can be asked to find the furthest distance of approach of the spacecraft if it were travelling away from the Earth without the motors being turned on.

Common learner misconceptions

- The powers of ten involved in the expressions for R and $\frac{1}{R}$ can easily be misunderstood. Teachers may need to explain that when $R/10^6$ m = 242 then $R = 242 \times 10^6$ m.

Answers to the workbook questions

a Gradient = $2GM$; intercept = $\frac{2E}{m}$

b See Table 10.4.

R / 10^6 m	v / m s^{-1}	$\frac{1}{R}$ / 10^{-8} m^{-1}	v^2 / 10^7 m^2 s^{-2}
242	1520	0.414	0.23
95.2	2720	1.05	0.74
54.3	3700	1.84	1.37
25.6	5490	3.91	3.01
13.3	7670	7.52	5.88
10.0	8850	10.0	7.83

Table 10.4

c See Figure 10.4.

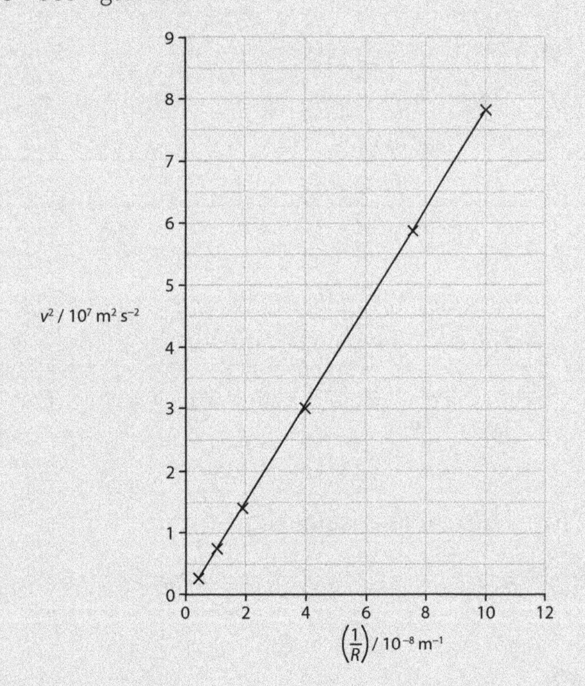

Figure 10.4

d Gradient = 0.793×10^{15}; unit = m^3 s^{-2}

e Mass of the Earth = $\frac{\text{gradient}}{2G} = 5.94 \times 10^{24}$ kg

Unit = $\frac{m^3\ s^{-2}}{N\ m^2\ kg^{-2}} = \frac{kg^2\ m\ s^{-2}}{N} = kg$

f intercept = $-0.1 \times 10^7 = 1.0 \times 10^6$ m^2 s^{-2}

g From the result in part **f**, $E = m \times \frac{\text{gradient}}{2}$

$= -3.0 \times 10^9$ J

The potential energy is negative and at any point, more negative than the kinetic energy. The spacecraft does not have sufficient kinetic energy to escape the pull of the Earth.

Chapter 11: Oscillations and communications

Chapter outline

This chapter relates to Chapter 19: Oscillations and Chapter 20: Communication Systems, in the coursebook.

In this chapter learners will complete investigations on:

- 11.1 The period of oscillation of a steel blade
- 11.2 Planning investigation into damped oscillations
- 11.3 Simple harmonic oscillation of a mass on a spring
- 11.4 Data analysis investigation into attenuation of a coaxial cable

Practical Investigation 11.1: The period of oscillation of a steel blade

Skills focus

See the Skills grids at the front of this book for details of the skills developed and used in this investigation.

Duration

The practical work will take 30 minutes; the analysis and evaluation questions will take 30 minutes.

Preparing for the investigation

Learners should know:

- the equation and units for Young modulus
- how to time oscillations using a fiducial mark placed at the centre of the oscillation
- the meaning of one complete oscillation
- how to combine percentage uncertainties from several readings of the same quantity
- how to plot error bars and use them to draw lines of best fit and worst fit.

You should put the masses on the end of the hacksaw blade and show learners how to clamp the blade without damage to the bench.

Learners will vary the distance *d* and measure the period of oscillation of the blade when it oscillates upwards and downwards with small amplitude.

Equipment

Each learner or group will need:

- steel blade, e.g. a hacksaw blade of about 30 cm in length or other similar steel blade of the same thickness and width. A piece of tape should be placed over the serrated edge of the blade so learners do not cut their hands when touching the blade
- stopwatch to read to a precision of at least 0.1 s
- two 50 g masses (or one 100 g mass); The masses should be securely attached to the end of the hacksaw blade with tape or modelling clay.
- G-clamp and small blocks of wood
- 30 cm ruler or metre rule.

Access to:

- micrometer screw gauge or callipers.

Safety considerations

- The masses should be securely fixed to the end of the blade so that when the blade oscillates they do not fall off.
- Any sharp edges on the blade should be taped over.

Carrying out the investigation

- The oscillation may be rather fast and placing a piece of paper gently against the oscillating blade allows one to hear when the blade reaches the end of an oscillation. This is particularly useful when the period is small.

Plotting logarithmic graphs that have negative values can prove difficult and teachers may need to give help. In particular, the choice of axes is important as the graph should **not** start at (0, 0).

Able learners can be challenged to use base units in the SI system to find a value for n with the equation:

$$T = d^n \times \sqrt{\frac{16\pi^2 M}{Ewt^3}}$$

See Table 11.1 for an example.

Quantity	E	M	w, d and t
Base units	$\text{kg m}^{-1}\text{s}^{-2}$	kg	m

Table 11.1

As the base units must be the same on each side of the equation, $n = 1.5$.

Common learner misconceptions

- Learners often think that one oscillation is from one extreme to the other.

Sample results

These should give an idea of the results the learners should end the investigation with.

d / m	T_{10} / s 1st	T_{10} / s 2nd	T_{10} / s average	T / s	lg (d / m)	lg (T / s)
0.125	2.16	2.26	2.21	0.221 ± 0.005	−0.903	−0.656 ± 0.010
0.145	2.52	2.68	2.60	0.26 ± 0.004	−0.839	−0.585 ± 0.013
0.173	3.50	3.72	3.61	0.361 ± 0.011	−0.762	−0.442 ± 0.013
0.216	4.53	4.95	4.74	0.474 ± 0.021	−0.666	−0.324 ± 0.019
0.242	5.60	5.85	5.73	0.573 ± 0.010	−0.616	−0.242 ± 0.009
0.276	6.86	6.36	6.61	0.661 ± 0.025	−0.559	−0.180 ± 0.016

Table 11.2

Width of blade $w = 0.0127$ m   Thickness of blade $t = 0.00080$ m

Answers to the workbook questions (using the sample results)

a See Table 11.2 for sample results

b $\lg T = \lg k + n \lg d$

c Gradient = n Intercept = $\lg k$

d See Table 11.2 for sample results

e, f See Figure 11.1.

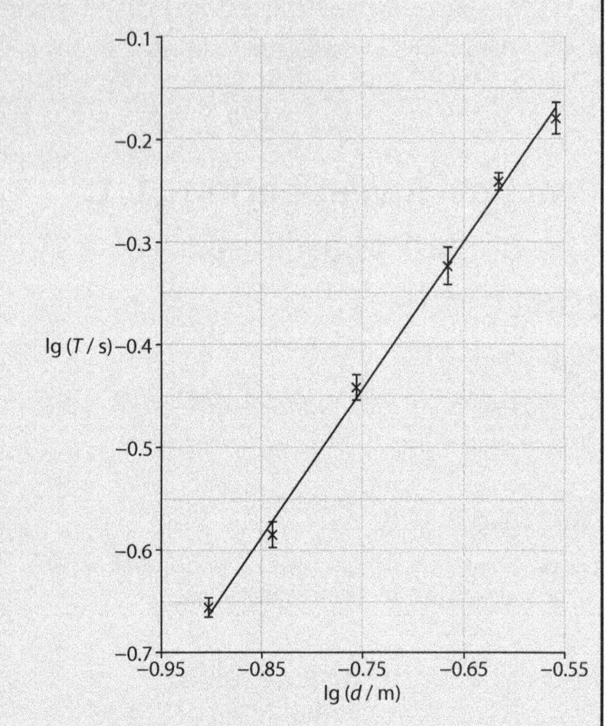

Figure 11.1

g Gradient of best-fit line = 1.42, gradient of worst-fit line = 1.44, uncertainty ±0.02

h $n = 1.42$, Uncertainty in $n = \pm 0.02$

i Point on graph, for example, $\lg(T / \text{s}) = -0.3$;
$\lg(d / \text{m}) = -0.65$
$\lg k = 0.623$, $k = 4.2$

j $E = 1.4 \times 10^{11}\ \text{N m}^{-2}$

Practical Investigation 11.2: Planning

Investigation into damped oscillations

Skills focus

See the Skills grids at the front of this book for details of the skills developed and used in this investigation.

Duration

The planning investigation will take about 40 minutes; carrying out the experiment after planning will take about 40 minutes.

Preparing for the investigation

It may be helpful to demonstrate a swinging simple pendulum with small piece of card attached.

Learners need to:

- know what is meant by the amplitude of an oscillation
- know about different types of variables and the control of quantities
- be able to handle natural logarithms and know the relationship: $\ln a^b = b \ln a$.

Learners will plan an experiment to show the exponential decay of amplitude A with the number of swings n. If apparatus is available then readings can be taken and the constant λ in the equation $A = A_0 e^{-\lambda n}$ can be determined. A sample set of readings is also provided.

Additional notes and advice

- There is no practical equipment necessary in the planning, although teachers may like to show a demonstration of a simple pendulum with a piece of card or paper at the bottom to increase damping. If the experiment is actually performed then items such as a mass on a string, metre rule and some pins on stands to mark the position of the mass can be used. Alternatively, a longer string can be used and marks made on the floor with the eye placed directly above the mass. If pins are used then learners must be instructed to be careful and it is sensible to use the flat end of a pin uppermost to avoid this problem.

Variables

Learners should identify:

- dependent variable: amplitude A
- independent variable: number of swings n
- variables to be controlled: mass, length of string, amount of air resistance.

These dependent and independent variables can be interchanged, leading to a slightly different method.

Equipment

Examples might be mass on string or metre rule with hole and pin to act as a pendulum, retort stand, boss and clamp and means of securing stand to the bench, such as a heavy weight or a G-clamp, long pins or nails and sticky tack to stick pins to the bench to mark the amplitude and a ruler (see Figure 11.2 for an example).

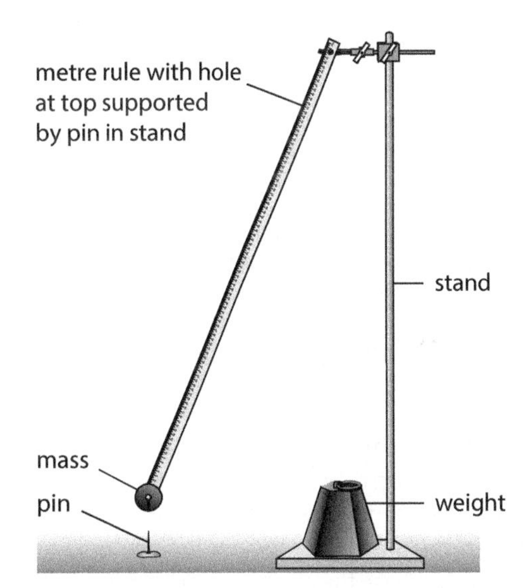

Figure 11.2

Alternatively, a camera and video recorder can be placed at right angles to the plane of oscillation of the pendulum with a metre rule alongside the oscillating mass.

Safety considerations

For example, **one** of the following:

- Ensure the stand does not fall over by clamping to the end with a G-clamp or adding a heavy weight.
- Take care using the pins.
- Ensure heavy weights do not fall over the bench onto your foot.

Method

A suggested method with the amplitude as the dependent variable is:

1 Start with the mass stationary at the middle of the oscillation and place a pin just under the centre of the mass. Move the mass to the starting point and hold it stationary. Place another pin just under the centre of the mass.

2 Release the mass and, after five swings, place a pin just under the mass at the furthest point it reaches from the middle of the oscillation and continue placing pins under the mass after 10, 15, 20, 25 and 30 swings.

3 Measure the distance from each pin to the pin at the middle of the oscillation. Repeat and average the results. The first result at the start is tabulated at $n = 0$.

4 To increase the air resistance add a piece of card to the mass at right angles to the plane of oscillation.

Some extra detail might be:

- Look directly at right angles to the plane of oscillation of the pendulum to avoid parallax error when judging the position of the pin, the height of which can be adjusted to just touch the mass.
- Five oscillations might not be enough. Place new pins when the oscillation has changed significantly and count the total number of oscillations each time.
- Make sure the ruler is horizontal as each measurement is made.
- When adding the card it may affect the period which is a control variable. The card may be added to the mass at the start of the experiment. In the beginning, the card is angled to be in the same plane as the oscillation of the pendulum, and to increase the air resistance the card is rotated by 90°

Carrying out the investigation

Learners who are not studying mathematics may need help in dealing with the logarithm of an exponential quantity.

Some learners, who have finished the investigation, can help others who may be struggling. However, this exercise provides the opportunity for such learners to take measurements and start the experiment, with others who have yet to finish joining them later. As an alternative, the sample measurements in Table 11.3 below can be analysed. The graph is shown in Figure 11.3 and λ is 0.024.

- If learners look at one another's plans and discuss them, they can make sensible comments to each other about whether the plan is logical. Such discussion is often very useful in demonstrating to each other which elements of a plan are missing.

Common learner misconceptions

- Learners may think that the distance between the two extremes of the motion is the amplitude. Although this distance can be measured and used, it is **not** the amplitude. Amplitude is measured to the **centre** of the oscillation.

Sample results

Table 11.3 provides sample results that learners may analyse.

Number of swings	Amplitude / cm	ln (A / cm)
0	35.2	3.561
5	30.8	3.428
10	26.8	3.288
15	23.2	3.144
20	20.8	3.035
25	18.8	2.934
30	16.8	2.821
35	14.8	2.695

Table 11.3

Answers to the workbook questions (using the sample results)

a Calculate values of $\ln A$ (or $\lg A$). Plot a graph of $\ln A$ on the y-axis and n on the x-axis. An example is shown in Figure 11.3.

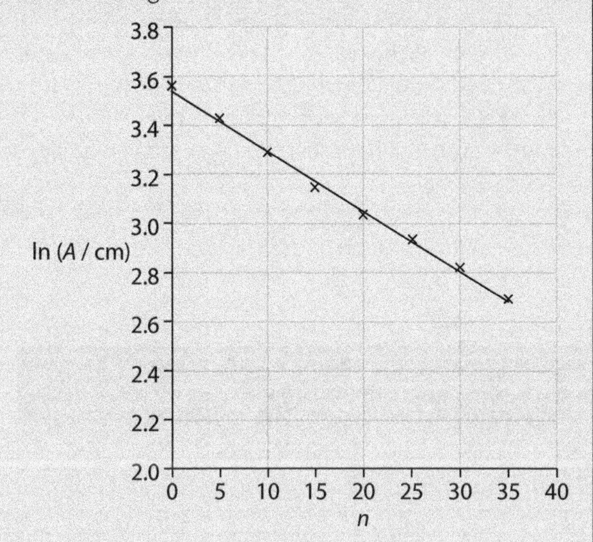

Figure 11.3

b Since $\ln A = \ln A_0 - \lambda t$ then the gradient of the graph is $-\lambda$.

c In the light of performing the experiment, learners may, for example, realise the necessity of avoiding parallax error either when placing pins or when measuring the distance. It is helpful if they have appreciated some problems for themselves and you can discuss their suggestions afterwards with the whole class. The particular problems and improvements will depend on the apparatus and methods chosen.

Practical Investigation 11.3: Simple harmonic oscillation of a mass on a spring

Skills focus

See the Skills grids at the front of this book for details of the skills developed and used in this investigation.

Duration

The practical work will take 30 minutes; the analysis and evaluation questions will take about 30 minutes.

Preparing for the investigation

Learners will need to:

- know how to time 10 complete oscillations using a fiducial mark such as a point on the retort stand near the centre of the oscillation
- know how to choose a sensible number of oscillations
- measure at least 10 oscillations with a stopwatch and write down the reading to 0.01 s, even though their reaction time is larger than this.
- be able to use logarithms and antilogarithms to base 10 and to use the relationships:

$$\lg(ab) = \lg a + \lg b \text{ and } \lg(a^b) = b \lg a$$

- measure the time for a number of oscillations for different masses hanging on the spring.

Equipment

Each learner or group will need:

- retort stand, boss and clamp
- G-clamp or heavy weight to stop the stand from toppling
- steel spring
- mass hanger 100 g with five 100 g slotted masses
- stopwatch.

Alternative equipment

Any spring or combination of springs, for example in series, can be used with any sensible range of masses as long as the time for 10 oscillations can be measured.

Safety considerations

- Make sure that the retort stand cannot topple. This can be done by placing a heavy weight on the base of the retort stand or by clamping the stand to the bench.

Carrying out the investigation

- The amplitude of the oscillation should not become too small in 10 oscillations and should not be so large that the tension in the spring reduces to zero.

  Learners may have difficulty in plotting and understanding a logarithmic graph, particularly with the negative values.

  More able learners can extend the investigation by connecting two springs in series or parallel and predicting the effect on the investigation and then obtaining readings to see whether their prediction is correct. For example, how does the period of the mass depend on the number n of identical springs in series?

They can also show by the method of dimensions that the constant b should be 0.5.

Common learner misconceptions

- Learners might think that they are timing 10 oscillations when they are only timing 9 oscillations.
- Measuring the time for 10 oscillations is best by counting 0 as the mass passes through the mid-position and starting the stopwatch. The stopwatch is stopped when the count has reached 10. Learners often think that they can obtain a more accurate period by measuring from one extreme of the oscillation rather than from the centre of the oscillation. As the mass is moving faster in the middle it is easier to time from this point. Learners often think that they should record the timings to the nearest 0.2 s as that is about their reaction time. Strictly, the uncertainty is the difference between the starting and stopping reaction times which may be less than 0.1 s but there is some merit in writing down the reading on any meter, even when the last digit may not be significant.

Sample results

The sample results in Table 11.4 can be used.

		t_{10} / s						
M / kg	n	1st	2nd	3rd	average	T / s	lg (T / s)	lg (M / kg)
0.100	10	6.92	6.89	6.95	6.92 ± 0.03	0.692 ± 0.003	−0.160 ± 0.002	−1.000
0.200	10	9.95	9.80	10.13	9.96 ± 0.12	0.996 ± 0.012	−0.002 ± 0.001	−0.699
0.300	10	12.10	11.99	12.25	12.11 ± 0.13	1.211 ± 0.013	0.083 ± 0.004	−0.523
0.400	10	14.00	13.90	14.10	14.00 ± 0.05	1.400 ± 0.005	0.146 ± 0.002	−0.398
0.500	10	15.28	15.22	15.23	15.24 ± 0.04	1.524 ± 0.004	0.183 ± 0.001	−0.301

Table 11.4

Answers to the workbook questions (using the sample results)

a See Table 11.4

b See Figure 11.4

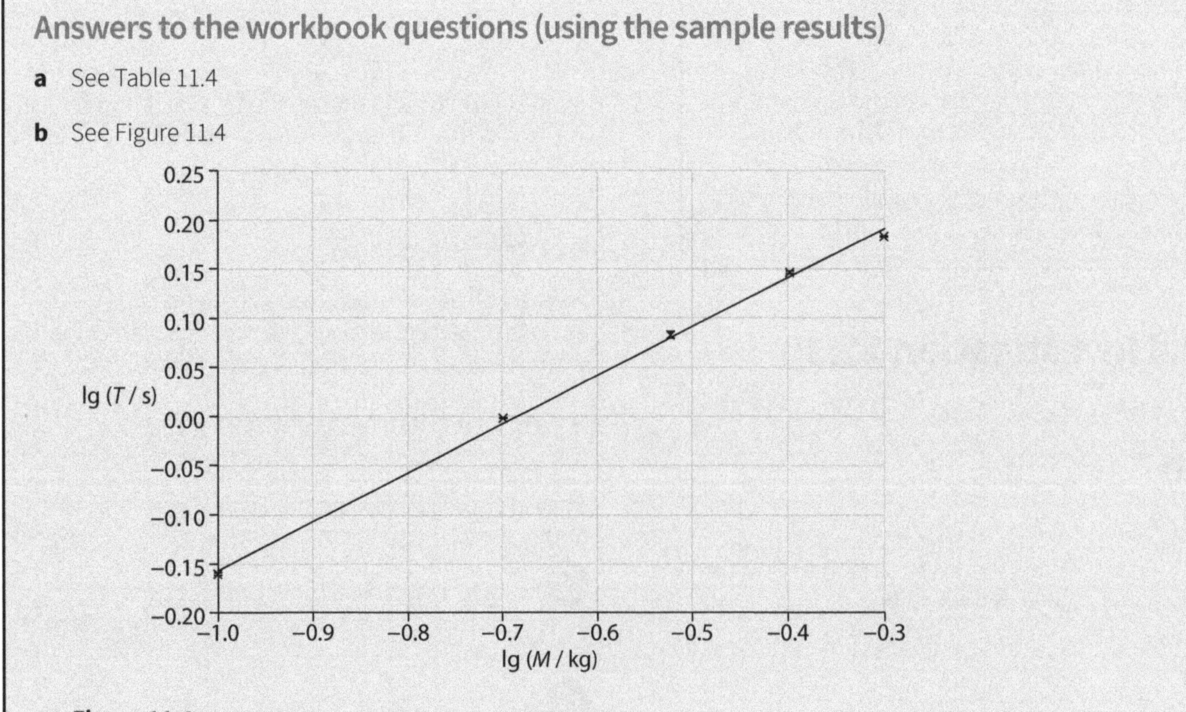

Figure 11.4

c Gradient = 0.495 ± 0.003 approximately

d $\lg T = \lg a + b \lg M$

e 0.495 ± 0.003 approximately

f Point on graph: y-value ($\lg T$) = 0.144; x-value ($\lg M$) = −0.400

$\lg a = 0.342$ $a = 2.20$

g The value for b should be 0.5 and this is only slightly outside the calculated uncertainty and so there is fair agreement.

h $k = 8.2\ (\text{N m}^{-1})$

Practical Investigation 11.4: Data analysis

Investigation into attenuation of a coaxial cable

Skills focus

See the Skills grids at the front of this book for details of the skills developed and used in this investigation.

Duration

The analysis and evaluation questions will take about 45 minutes.

Preparing for the investigation

- It may be helpful if learners have met the idea of attenuation and can calculate values of attenuation in dB.
- Learners need to be able to use logarithms to base 10 and draw graphs and find gradients, including worst-fit lines.
- Learners will be calculating the attenuation of a coaxial cable for different frequencies.

Carrying out the investigation

Learners may need help in plotting error bars when the frequency value is known to within 10%.

More able learners can be asked to calculate the attenuation per unit length of the cable and given problems that involve this figure. They may also be asked to plan an investigation using the same equipment to determine the relationship between the attenuation of the cable and its length at a given frequency.

Answers to the workbook questions

a Independent variable: frequency

Dependent variable: power output or attenuation

Variables that are controlled: length of coaxial cable, input power, load resistor

b, c See Table 11.5.

f / MHz	P_{in} / W	P_{out} / W	Attenuation / dB	Attenuation^2 / dB^2
10	0.001800	0.001427	1.01	1.02
55	0.001800	0.001020	2.47	6.10
211	0.001800	0.000608	4.71	22.2
250	0.001800	0.000530	5.31	28.2
300	0.001800	0.000462	5.91	34.9

Table 11.5

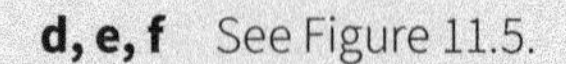

d, e, f See Figure 11.5.

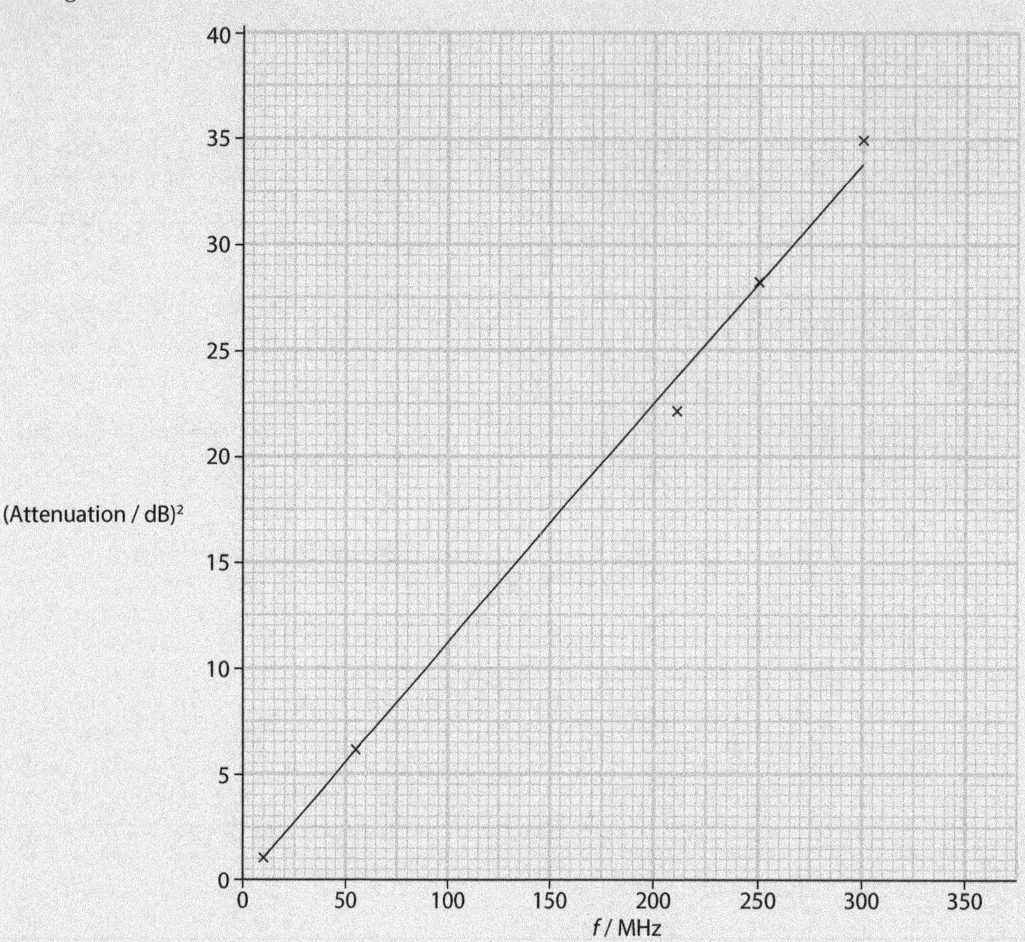

Figure 11.5

g Gradient line of best fit = 0.105, gradient line of worst fit = 0.110

Gradient = $0.105 \pm 0.005\ (\text{MHz})^{-1}$

h Gradient = k^2

i Yes it does, because the graph of attenuation² is proportional to frequency as it is a straight line through the origin within the limits of the uncertainty.

Chapter 12:
Thermal physics and ideal gases

Chapter outline

This chapter relates to Chapters 21: Thermal physics and Chapter 22: Ideal gases, in the coursebook.

In this chapter learners will complete investigations on:

- 12.1 Data analysis investigation into the thermocouple
- 12.2 Boyle's law
- 12.3 Planning investigation into specific latent heat of vaporisation of water
- 12.4 Data analysis investigation into specific latent heat of vaporisation of water

Practical Investigation 12.1: Data analysis

Investigation into the thermocouple

Skills focus

See the Skills grids at the front of this book for details of the skills developed and used in this investigation.

Duration

The analysis and evaluation questions will take about 30 minutes. Any further practical investigation can take about 30 minutes.

Preparing for the investigation

- It may be helpful to explain that the e.m.f. of a thermocouple is just the voltage output and this is very low. You can demonstrate such a thermocouple simply, with two different metals connected to a multimeter on the mV scale. The voltage is very low and so a d.c. amplifier is required.
- Learners may not have studied electronics as yet and so no detail is required except, simply, that a d.c. voltage is multiplied by a constant factor, in this case 10.0. If enough multimeters and thermocouples are available then learners can investigate some of these values for themselves although the accuracy will be very limited as the precision of a multimeter is usually only to the nearest mV.
- The learners will investigate how the thermocouple e.m.f. varies with temperature.

Equipment

Each learner or group will need:

- a copper-constantan thermocouple (this is simply two copper wires and a wire of constantan joined in series with the constantan in the middle section; these can be just twisted together and may be soldered if it is wished to keep the thermocouple for future use)
- a digital millivoltmeter reading to a precision of at least 1 mV; a cheap multimeter is suitable
- a beaker
- access to hot water, e.g. a kettle and ice.

Safety considerations

- Learners should be warned about the danger of carrying hot liquid around the laboratory, both to themselves and to others, if any practical work is carried out.

Carrying out the investigation

A few learners may need to be shown how to find 0.5% of the e.m.f. to plot as the length of each error bar. One row has been completed in Table 12.1 in the workbook to help learners to understand and to check their calculation.

Learners who have completed the task could use a thermocouple and place it in hot water with a thermometer and compare the speed of response and perhaps find the temperature of the water using the equation for V they have obtained, if they have a copper constantan thermocouple and a millivoltmeter.

Common learner misconceptions

- Some learners may wonder why the value of the e.m.f. at 0 °C is not included and you may demonstrate that $\frac{0}{0}$ is not defined and can have any value.

Answer to workbook questions

a $\frac{V}{T} = a + bT$

b Gradient = b and intercept = a

c See Table 12.1.

Temperature T / °C	Voltmeter reading / mV	E.m.f. of thermo-couple / mV	$\frac{V}{T}$ / mV °C^{-1}
10	3.9	0.39	0.0390 ± 0.002
20	7.9	0.79	0.0395 ± 0.002
30	12.0	1.20	0.0400 ± 0.002
40	16.1	1.61	0.0403 ± 0.002
50	20.4	2.04	0.0408 ± 0.002
60	24.7	2.47	0.0412 ± 0.002
70	29.1	2.91	0.0416 ± 0.002
80	33.6	3.36	0.0420 ± 0.002
90	38.1	3.81	0.0423 ± 0.002
100	42.8	4.28	0.0428 ± 0.002

Table 12.1

d See Figure 12.1

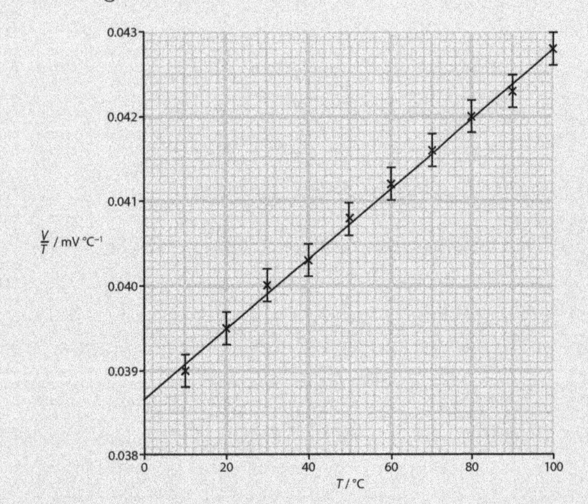

Figure 12.1

e Gradient of best-fit line = 4.0×10^{-5}, worst-fit line about 4.6×10^{-5} mV °C^{-2}

Gradient of best-fit line = $4.0 \times 10^{-5} \pm 0.6 \times 10^{-5}$ mV °C^{-2}

Intercept of best-fit line = 0.0387; worst-fit line about 0.0384 mV °C^{-1}

Intercept of best-fit line = 0.0387 ± 0.003 mV °C^{-1}

f $a = 0.0387 \pm 0.003$ mV °C^{-1},
$b = 4.0 \times 10^{-5} \pm 0.6 \times 10^{-5}$ mV °C^{-2}

g Neither thermometer has a change in thermometric property that is proportional to temperature, which the temperature scale of the particular thermometer assumes. The two thermometers will agree at the fixed points but not accurately at any other temperature.

Practical Investigation 12.2: Boyle's law

Skills focus

See the Skills grids at the front of this book for details of the skills developed and used in this investigation.

Duration

The practical work will take 40 minutes; the analysis and evaluation questions will take about 30 minutes.

Preparing for the investigation

Learners should have some understanding of the gas laws and of the gas constant in particular.

They should be able to:

- convert temperatures into K
- determine uncertainty from several readings of the same quantity
- plot error bars and use them to draw lines of best fit and worst fit.

Learners will hang masses from the bottom of the piston of the syringe and measure the volume of the gas in the syringe.

Equipment

Each learner or group will need:

- 10 ml disposable syringe, sealed at one end. This is best done by placing glue in one end or sealing it by heating. Before sealing, the plunger should be adjusted so that there is about 6 ml of air in the syringe.
- stand and clamp to hold the syringe. The syringe should be clamped upside down so that the plunger can move down freely; allow sufficient distance above the bench for the mass hanger plus about 15 cm of free travel for the plunger to move down.
- 30 cm ruler
- 100 g slotted mass hanger and up to nine 100 g slotted masses
- loop of string which is tied to the handle of the syringe so that the slotted mass hanger can be suspended from it.

Access to:

- a thermometer to measure room temperature.

Safety considerations

- You should ensure that the syringe is supported firmly upside down by the clamp and stand and that the loop of string is tied securely round the handle of the syringe before the start of the experiment. Learners will hang 1 kg onto the string attached to the plunger and the outside of the syringe must be firmly supported in order not to fall down.

Carrying out the investigation

- Learners should repeat each reading at least once. The plunger may stick as it moves up and down the tube and a little grease may be needed in order to reduce this effect. Learners should push the plunger up and down before each reading.

  Learners may need to be given or shown where to find the value of the gas constant and reminded how to convert temperatures from °C to K.

  Learners may be given the task of showing that the units of the equation used in part **f** to find the number of moles actually produces the unit moles, as long as the graph is drawn with the *y*-axis having units m^{-1}. They should also be encouraged to apply the gas laws to obtain the original equation for themselves.

Common learner misconceptions

- Learners should make sure that they give a negative value for the gradient and that this sign is not lost.
- Learners may believe that there is no uncertainty if their two values for the length are the same. Teachers may need to suggest that the smallest scale reading on the ruler is the uncertainty in this case.

Sample results

The results in Table 12.2 give an idea of the results the learners should produce in the investigation.

	$\frac{l}{m}$			
***M* / kg**	**1st reading**	**2nd reading**	**average and uncertainty**	$\frac{1}{l}$ **/ m^{-1}**
0.000	3.0×10^{-2}	3.0×10^{-2}	$3.0 (\pm 0.1) \times 10^{-2}$	33 ± 1
0.200	3.3×10^{-2}	3.2×10^{-2}	$3.3 (\pm 0.1) \times 10^{-2}$	31 ± 9
0.400	3.7×10^{-2}	3.5×10^{-2}	$3.6 (\pm 0.1) \times 10^{-2}$	28 ± 7
0.600	4.2×10^{-2}	4.0×10^{-2}	$4.1 (\pm 0.1) \times 10^{-2}$	24 ± 6
0.800	4.7×10^{-2}	4.3×10^{-2}	$4.5 (\pm 0.2) \times 10^{-2}$	22 ± 10
1.000	5.6×10^{-2}	5.2×10^{-2}	$5.4 (\pm .2) \times 10^{-2}$	19 ± 7

Table 12.2

Answers to the workbook questions (using the sample results)

a Gradient $= -\frac{g}{nRT}$ and intercept $= \frac{p_0 A}{nRT}$

b See Table 12.2.

c See Figure 12.2

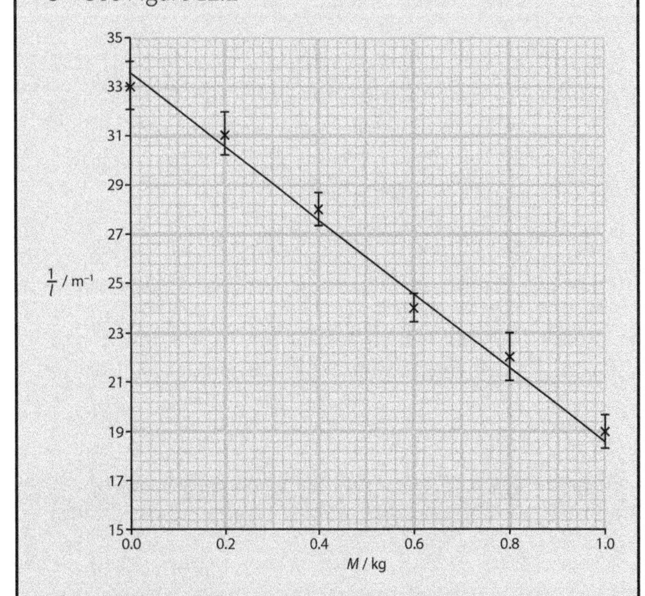

Figure 12.2

d Gradient of line of best fit = −14.7, gradient of line of worst fit = −16 approximately

Gradient $= -15 \pm 1$ $kg^{-1} m^{-1}$

e Room temperature = 20 °C = 293 K

f Number of moles $n = \frac{g}{\text{gradient} \times RT} = \frac{9.81}{15 \times 8.31 \times 293} = 2.7 \times 10^{-4} \pm 0.2 \times 10^{-4}$

g $p_0 = \text{intercept} \times \frac{nRT}{A}$

h The gradient would have a value twice as negative and the intercept would be doubled as the number of moles is halved.

Practical Investigation 12.3: Planning

Investigation into specific latent heat of vaporisation of water

Skills focus

See the Skills grids at the front of this book for details of the skills developed and used in this investigation.

Duration

The planning investigation will take about 40 minutes.

Preparing for the investigation

It may be helpful to show a simple apparatus used to boil water, preferably using a low voltage. An electric kettle can even be used, although this cannot readily be used to take readings as varying and measuring the power input is not easy because the voltage is mains voltage.

Learners need to:

- know the formulae for specific latent heat and its units
- know the formula for electrical power and how to connect both an ammeter and a voltmeter
- be able to draw graphs and calculate their gradient.

Learners will plan how to vary the power into a heater and measure how this changes the mass of water evaporated.

Additional notes and advice

- There is no practical equipment necessary for learners to use. If a demonstration is shown to learners then a suitable heater for heating a cup of tea from a car battery may be used (about 1–2 Ω) and a low voltage supply (up to 12 V). Multimeters, measuring up to 10 A and 12 V are suitable and a top-pan balance will be required.

Variables

Learners should identify:

- dependent variable: mass of water m that turns to steam
- independent variable: power of heater P *or* V and I
- variables to be controlled: time of heating t, same resistor (or heater), and any variable that may affect the heat loss; for example, the surface area, thickness of the cup, draughts or, alternatively, the amount of the heater that is exposed to the air.

Equipment

Each learner or group will need:

- voltmeter and ammeter (or joulemeter)
- stopwatch
- heater
- beaker
- low voltage power supply
- ammeter
- voltmeter
- top-pan balance.

Safety considerations

- If any practical work is carried out then handling of boiling water must be done safely, with the cup held and transported safely and low voltages used.
- Gloves may be used to handle the cup or any sensible safety precaution suggested.

Method

A suggested method is as follows.

1 Set up the apparatus as shown in Figure 12.3. Turn the heater on and wait until the water is boiling steadily.

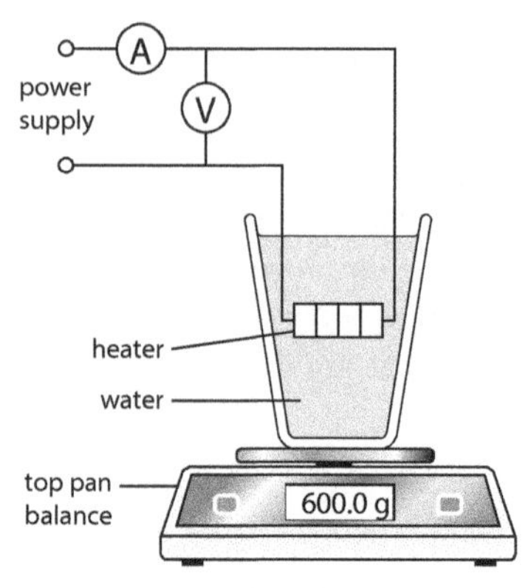

Figure 12.3

2 Start the stopwatch and, at the same instant, record the reading on the top-pan balance.

3 Record the reading on the ammeter and voltmeter regularly as the water boils.

4 After 5 minutes record the new reading on the top-pan balance.

5 Top up the cup until there is the same initial level of water in the cup. Repeat the procedure above.

6 Change the voltage supplied by the power supply and repeat the procedure above.

Extra detail, for example:

- adjust variable power supply / variable resistor to ensure p.d. / current is constant
- more detail about how to keep heat loss constant, e.g. add insulation to sides of beaker
- use large time to reduce percentage uncertainty
- relationship is valid if the graph is a straight line through the origin with a negative intercept or zero intercept if H is negligible.

Results

Table 12.3 shows what the results table might look like.

Voltage / V	1st		
	2nd		
	average		
Current / A	1st		
	2nd		
	average		
Top-pan balance reading / g	initial		
	final		
	difference = m		

Table 12.3

Carrying out the investigation

Some learners may need help in understanding that the power is being varied and how this may be achieved, or in rearranging the initial formula to make m the subject of the formula.

Some learners, who have finished the investigation, can be asked to predict how the graph and equation will alter if the container for the water is made thinner or some other change is made.

Common learner misconceptions

- Learners may not realise that it is important to only start timing once the water is boiling at a steady rate.

Answers to the workbook questions

a Power is calculated using the formula $P = VI$, using the average values of voltage V and current I.

The mass m of water evaporated in 5 minutes is calculated by subtracting the final reading of the top-pan balance from the initial reading.

A graph of P on the x-axis is drawn against m on the y-axis. The gradient of the line is measured and the gradient $= \frac{t}{L}$.

The specific heat of vaporisation L is found by calculating $\frac{t}{\text{gradient}} = \frac{300}{\text{gradient}}$.

b One arrangement to measure the latent heat of fusion is for the heater to be placed in crushed ice placed inside a funnel. The water that melts in 300 s can be collected in a beaker and its mass measured, once the ice is melting at a constant rate. Since the latent heat of fusion is much less than the latent heat of vaporisation the gradient will be less and the value of H will be negative as heat is transferred from the environment.

Practical Investigation 12.4: Data analysis

Investigation into specific latent heat of vaporisation of water

Skills focus

See the Skills grids at the front of this book for details of the skills developed and used in this investigation.

Duration

The analysis and evaluation questions will take about 40 minutes.

Preparing for the investigation

It may be helpful to show a simple apparatus used to boil water, preferably using a low voltage. An electric kettle can even be used. However, this cannot readily be used to take readings as varying and measuring the power input is not easy as the voltage is mains voltage.

Learners need to:

- know the formulae for specific latent heat and its units
- know the formula for electrical power and how to connect both an ammeter and a voltmeter
- be able to draw graphs and calculate their gradient
- be able to combine uncertainties.

Learners will be using data on the input power to a heater and the mass of water evaporated to measure the specific latent heat of vaporisation and the heat loss from a cup in 3 minutes.

Additional notes and advice

- There is no practical equipment necessary for learners to use. If a demonstration is shown to learners then a suitable heater for heating a cup of tea from a car battery may be used (about 1–2 Ω) and a low voltage supply (up to 12 V). Multimeters, measuring up to 10 A and 12 V are suitable and a top-pan balance will be required.
- If any practical work is carried out then handling of boiling water must be done safely, with the cup held and transported safely and low voltages used.

Answers to the workbook questions

a $m = \frac{Pt}{L} - \frac{H}{L}$, gradient $= \frac{t}{L}$, intercept $= -\frac{H}{L}$

b See Table 12.4.

Voltage / V	Current / A	Power / W	*m* / g 1st	2nd	average	uncertainty
6.4	5.1	32.6	1.9	2.5	2.2	±0.3
7.5	6.1	45.8	2.6	3.1	2.9	±0.3
8.4	6.9	58.0	3.8	4.2	4.0	±0.2
8.6	6.9	59.3	3.9	4.2	4.1	±0.2
8.9	7.2	64.1	4.5	4.9	4.7	±0.2

Table 12.4

c, d See Figure 12.4

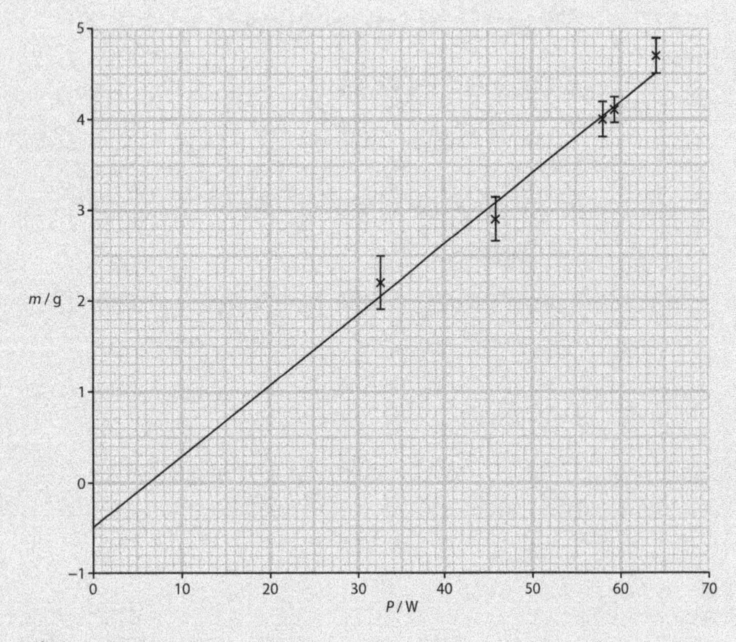

Figure 12.4

e Gradient of line of best fit = 0.078
line of worst fit = 0.083 approximately
gradient = 0.078 (±0.005) ($g\,W^{-1}$)

f Intercept = −0.50 g

g $L = \frac{180}{\text{gradient}} = 2300\ J\,g^{-1}$, $H_0 = -\text{intercept} \times L$
$= 0.50 \times 2300 = 1150\ J$

h % uncertainty in L = % uncertainty in gradient + % uncertainty in t = 6.4 + 1.1 = 7 or 8%

i Otherwise some of the heat energy is used to warm the water to its boiling point.

j The gradient is doubled but also the value of H, the thermal energy lost in time t, is doubled and so the intercept is twice as low.

Coulomb's law, capacitance and electronics

Chapter outline

This chapter relates to Chapter 23: Coulomb' law, Chapter 24: Capacitance and Chapter 25: Electronics, in the coursebook

In this chapter learners will complete investigations on:

- 13.1 Planning investigation into how the time for the potential difference across a capacitor to halve varies with the resistance
- 13.2 Determination of the capacitance of a capacitor in a d.c. circuit
- 13.3 Planning investigation into how the peak current in a capacitor circuit varies with the frequency of the a.c. supply
- 13.4 Determination of the capacitance of a capacitor in a a.c. circuit
- 13.5 Planning investigation into how the resistance of a thermistor varies with temperature
- 13.6 Investigation into an op-amp circuit.

Practical Investigation 13.1: Planning

Investigation into how the time for the potential difference across a capacitor to halve varies with the resistance

Skills focus

See the Skills grids at the front of this book for details of the skills developed and used in this investigation.

Duration

The planning investigation will take about 40 minutes.

Preparing for the investigation

It may be helpful to show a capacitor.

Learners need to:

- know about different types of variables and the control of quantities
- be able to draw graphs and calculate their gradient.

Variables

Learners should identify:

- The dependent variable is R, the resistance of the resistor.
- The independent variable is t, the time for the p.d. across the capacitor to halve.
- The variables to be controlled are keeping the capacitor constant. Learners may also suggest keeping the initial p.d. across the capacitor constant.

Equipment

Each learner or group will need:

- capacitor
- various resistors
- voltmeter
- power supply
- stopwatch
- connecting leads.

Safety considerations

- Low voltage power supplies should be used.
- If electrolytic capacitors are used, ensure that the polarity of the capacitor is correct.

Method

A suggested method is as follows.

- Labelled diagram showing a charging circuit for the capacitor and a voltmeter across the capacitor. The circuit should clearly be workable when the capacitor is discharging.

- The method should clearly indicate that the power supply is disconnected from the discharge circuit when the stop watch is started. This could be shown by a 'flying' lead.

Extra detail, for example:

- method to determine the value of each resistor
- care to connect the capacitor correctly to the power supply
- equation rewritten as an equation of a straight line, i.e. $t = -C\,ln\left(\frac{1}{2}\right)R = 0.693CR$.

Carrying out the investigation

- Learners may need help determining the logarithmic relationship.
- If new electrolytic capacitors are to be used to carry out the investigation after the planning stage, the manufacturer's guidance should be followed for the first use.

  Some learners will need help choosing an appropriate graph to plot.

  Learners who have finished the investigation could trial their plan and analyse their results. A possible capacitor to use would be 470 μF and resistors of the order of 105 Ω.

  Learners could derive the theoretical equation from the decay equation $V = V_0 e^{-\frac{t}{CR}}$.

  Learners could also compare with the radioactive decay equations.

Common learner misconceptions

- Learners may have difficulties in realising that there needs to be a charging circuit which needs to be disconnected for the discharge to occur.

Answers to the workbook questions

a Plot a graph of t against R

- Relationship is valid if the result is a straight line passing through the origin:

$$C = \frac{\text{gradient}}{0.693}$$

Practical Investigation 13.2: Determination of the capacitance of a capacitor in a d.c. circuit

Skills focus

See the Skills grids at the front of this book for details of the skills developed and used in this investigation.

Duration

The practical will take about 60 minutes.

Preparing for the investigation

See details provided in the workbook Practical investigation 13.2

Carrying out the investigation

- Learners may need help in calculating uncertainties in natural logarithms and the drawing of the worst acceptable line.
- Learners may also find it difficult to determine I_0 from the y-intercept.

  Learners may also need help in the mathematical analysis. Units may be problematical when determining both C and I_0.

  Learners who have finished the investigation could repeat the experiment with either a capacitor of a different value or a resistor of a different value.

  Learners could also determine the uncertainty in the value of capacitance assuming the resistance of the resistor had a tolerance of ± 5%.

Common learner misconceptions

- In the table of results, learners may determine logarithms to the base 10 rather than natural logarithms.
- Learners will need assistance in recording ln I to the correct number of significant figures.

Sample results

$R = 100\ k\Omega$

t / s	I / μA	ln (I / μA)
0	90±2	4.50±0.02
10	74±2	4.30±0.03
20	58±2	4.06±0.03
30	48±2	3.86±0.04
40	38±2	3.65±0.05
50	32±2	3.47±0.06
60	24±2	3.18±0.08
70	20±2	3.01±0.10

Table 13.1

Answers to the workbook questions (using the sample results)

a See Table 13.1.

b, e See Figure 13.1.

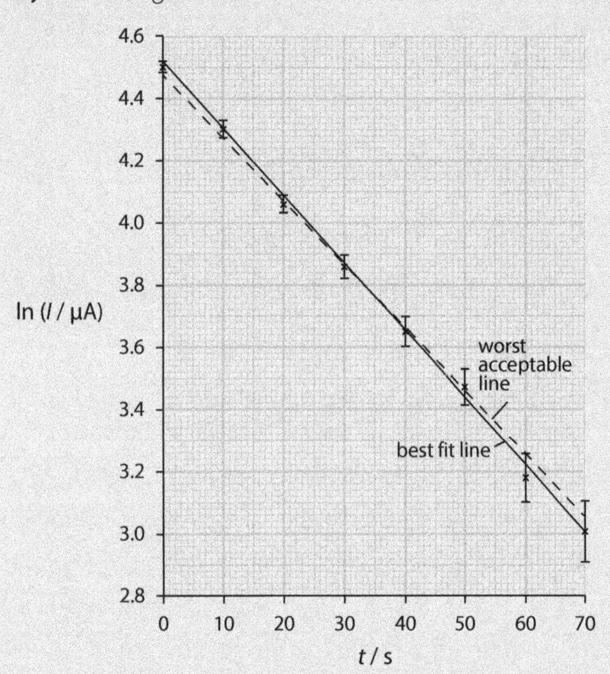

Figure 13.1

c $\ln I = -\frac{t}{CR} + \ln I_0$

d Gradient $= -\frac{1}{CR}$, y-intercept $= \ln I_0$

f Gradient = −0.0215; gradient of worst line = −0.0205 or −0.0225 Uncertainty in gradient = ±0.001

g y-intercept = 4.505; y-intercept of worst line = 4.52 or 4.49 Uncertainty in y-intercept = ±0.015

h $R = 100\,\text{k}\Omega$, $C = 470\,\mu\text{F}$, $I_0 = 90.5\,\mu\text{A}$

i Percentage uncertainty in C = ±4.7%
Percentage uncertainty in I_0 = ±2.5%

j Difficult to read the current at exactly the correct time. There is also an uncertainty in the value of R, which could be measured directly.

k If the current is read after the time has occurred, the currents should all be slightly larger. This means that the gradient will be slightly steeper. If the gradient has a larger magnitude, then the value of C will be smaller.

Practical Investigation 13.3: Planning

Investigation into how the peak current in a capacitor circuit varies with the frequency of the a.c. supply

Skills focus

See the Skills grids at the front of this book for details of the skills developed and used in this investigation.

Duration

The planning investigation will take about 40 minutes.

Preparing for the investigation

Learners need to:

- know about measuring frequency
- know about different types of variables and the control of quantities
- be able to draw graphs and calculate their gradient.

Variables

Learners should identify:

- The dependent variable is I_0, the peak current.
- The independent variable is f, the frequency of the alternating current.
- The variables to be controlled are keeping V_0 the peak voltage constant, the resistance of R constant and use the same capacitor.

Equipment

Each learner or group will need:

- capacitor
- resistor
- a.c ammeter or oscilloscope
- signal generator
- connecting leads.

Safety considerations

There are no specific safety considerations for this practical.

Method

A suggested method is as follows:

- labelled diagram showing a method for determining the peak current I_0
- the method could include either the use of an ammeter or the p.d. across the resistor.

Extra detail, for example:

- method to determine peak current I_0 from the ammeter in terms of I_{rms}.
- method to determine I_0 from an oscilloscope
- method to determine the frequency f of the signal generator using the oscilloscope
- method to measure V_0
- method to determine C
- equation re-written in a logarithmic form $\log I_0 = n \log f + \log(kCV_0)$.

Carrying out the investigation

- Learners may need help determining the logarithmic relationship.

  Some learners will need help choosing an appropriate graph to plot.

  Learners who have finished the investigation, could trial their plan and analyse their results: a possible capacitor to use would be 470 μF and resistors of the order of 100 Ω.

  Learners could determine k in terms of π and then write the relationship.

Common learner misconceptions

- Learners may have difficulties in measuring the peak current I_0.

Answers to the workbook questions

a Plot a graph of $\log I_0$ against $\log f$.

Relationship is valid if the result is a straight line.

$$n = \text{gradient}$$

$$k = \frac{10^{y\text{-intercept}}}{CV_0}$$

Practical Investigation 13.4: Determination of the capacitance of a capacitor in a a.c. circuit

Skills focus

See the Skills grids at the front of this book for details of the skills developed and used in this investigation.

Duration

The practical will take about 60 minutes.

Equipment

Each learner or group will need:

- capacitor
- 100 Ω resistor
- cathode ray oscilloscope
- signal generator
- connecting leads.

Safety considerations

There are no specific safety considerations for this practical.

Preparing for the investigation

See details provided in Practical investigation 13.4 in the workbook.

Carrying out the investigation

- Learners may need help in calculating uncertainties in $\frac{1}{f}$ and the drawing of the worst acceptable line.

  Learners may also need help in the mathematical analysis.

  Units may be problematical when determining C.

  Learners may need help to determine the uncertainty in X.

  Learners may need help in determining the power of ten in their final answers.

  Some learners will need to be reminded of the need to include a quantity and correct unit for the column headings in all their tables.

  Learners who have finished the investigation, could repeat the experiment with either a capacitor of a different value or a resistor of a different value. Learners could also determine the uncertainty in the value of capacitance assuming the resistance of the resistor has a tolerance of ±5%.

Common learner misconceptions

- In Table 13.1, significant figures may need to be reviewed.
- Some learners will need to be reminded of the need to include the quantity and correct unit for the column headings $\frac{1}{f}$, I_0 and X.

Sample results

$V_0 = 0.50 \pm 0.02$ V and $R = 100\ \Omega$

f / Hz	$\frac{1}{f}$ / 10^{-3} Hz^{-1}	V_{max} / V	I_0 / A	X / Ω
250	4.00	1.72±0.02	0.0172±0.0002	29.1±1.5
350	2.86	2.42±0.02	0.0242±0.0002	20.7±1.0
450	2.22	3.12±0.02	0.0312±0.0002	16.0±0.7
550	1.82	3.80±0.02	0.0380±0.0002	13.2±0.5
650	1.54	4.50±0.02	0.0450±0.0002	11.1±0.5
750	1.33	5.18±0.02	0.0518±0.0002	9.65±0.4

Table 13.2

Answers to the workbook questions (using the sample results)

a, b $V_0 = 0.50 \pm 0.02$V

See Table 13.2

c, e See Figure 13.2.

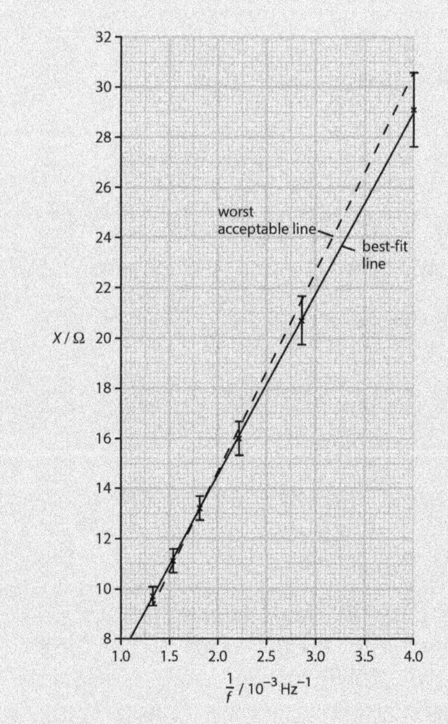

Figure 13.2

d Gradient $= \frac{1}{2\pi C}$

f Gradient = 7300

Gradient of worst-fit line = 6570 or 8010

Uncertainty in gradient = ±700

g $C = 22\ \mu$F

h Percentage uncertainty in C = 8%

i Absolute uncertainty in $C = \pm 2\ \mu$F

j Determine the frequency from the oscilloscope. Use the timebase to determine the period and then use the relationship $f = \frac{1}{T}$.

k Make sure that the waveform is as large as possible on the oscilloscope screen.

Practical Investigation 13.5: Planning

Investigation into how the resistance of a thermistor varies with temperature

Skills focus

See the Skills grids at the front of this book for details of the skills developed and used in this investigation.

Duration

The planning investigation will take about 40 minutes.

Preparing for the investigation

- It may be helpful to show a thermistor.

Learners need to:

- know about different types of variables and the control of quantities
- be able to draw graphs and calculate their gradient.

Variables

Learners should identify:

- the dependent variable is R, the resistance of the thermistor
- the independent variable is T, the absolute temperature of the thermistor

Equipment

Each learner or group will need:

- beaker, water
- thermometer
- thermistor or ohmmeter
- stirrer
- equipment to heat water
- power supply, ammeter and voltmeter to connect to the thermistor or ohmmeter.

Safety considerations

- Wear gloves when handling beaker and hot water.

Method

A suggested method is as follows:

- labelled diagram showing the basic arrangement of apparatus with a thermistor, a thermometer and a method of heating
- labelled diagram showingthe circuit (with correct circuit symbols) connected to the thermistor.

Extra detail, for example:

- stir the water
- wait for temperature to equalise
- insulate the beaker
- equation for determining resistance
- absolute temperature = temperature measured in °C + 273
- the given equation could be given in its logarithmic form, i.e. $\lg R = q \lg T + \lg p$.

Carrying out the investigation

- Learners may need help determining the logarithmic relationship.

  Some learners will need help choosing an appropriate graph to plot.

  Learners who have finished the investigation could trial their plan and analyse their results.

Common learner misconceptions

- Learners will often draw one circuit for both the light source and the LDR.
- Learners may need reminding of the significance of absolute temperature.

Answers to the workbook questions

a Plot a graph of $\lg R$ against $\lg T$

Relationship is valid if the result is a straight line with gradient q

q = gradient

$p = 10^{y\text{-intercept}}$

Practical Investigation 13.6: Investigation into an op-amp circuit

Skills focus

See the Skills grids at the front of this book for details of the skills developed and used in this investigation.

Duration

The practical will take about 60 minutes.

Preparing for the investigation

- See details provided in the workbook Practical investigation 13.6.

Equipment

Each learner or group will need:

- op-amp
- 10 000 Ω resistor
- 2700 Ω resistor
- six 1000 Ω resistors
- 3 V battery
- power supply for op-amp
- voltmeter (0–10 V)
- connecting leads.

Safety considerations

- Learners should take care when handling hot water.
- Learners should wear heatproof gloves.

Carrying out the investigation

- Learners may need help in connecting the power supplies to the op-amp and the general set-up of the circuit.

  Learners may also need help in the mathematical analysis and the drawing of the graph.

  Some learners will need to be reminded of the need to include a quantity and correct unit for the column headings in all their tables.

  Learners could also investigate the difference in uncertainty values when determining the uncertainty in *X* by using the gradient with the *y*-intercept as opposed using *E*, *R* and the *y*-intercept.

Common learner misconceptions

- Learners will need assistance in recording to the correct number of significant figures.
- The graph with negative values is difficult to draw.
- Some learners will need to be reminded of the need to include the quantity and correct unit for the column headings $\frac{1}{Y}$.

Sample results

Y / Ω	$\frac{1}{Y}$ / 10^{-4} Ω^{-1}	V / V
1000 ± 100	10.0 ± 1	−8.91
2000 ± 200	5.00 ± 0.51	−4.86
3000 ± 300	3.33 ± 0.34	−3.51
4000 ± 400	2.50 ± 0.25	−2.84
5000 ± 500	2.00 ± 0.20	−2.43
6000 ± 600	1.67 ± 0.17	−2.16

Table 13.3

Answers to the workbook questions (using the sample results)

a See Table 13.3

b, d See Figure 13.3

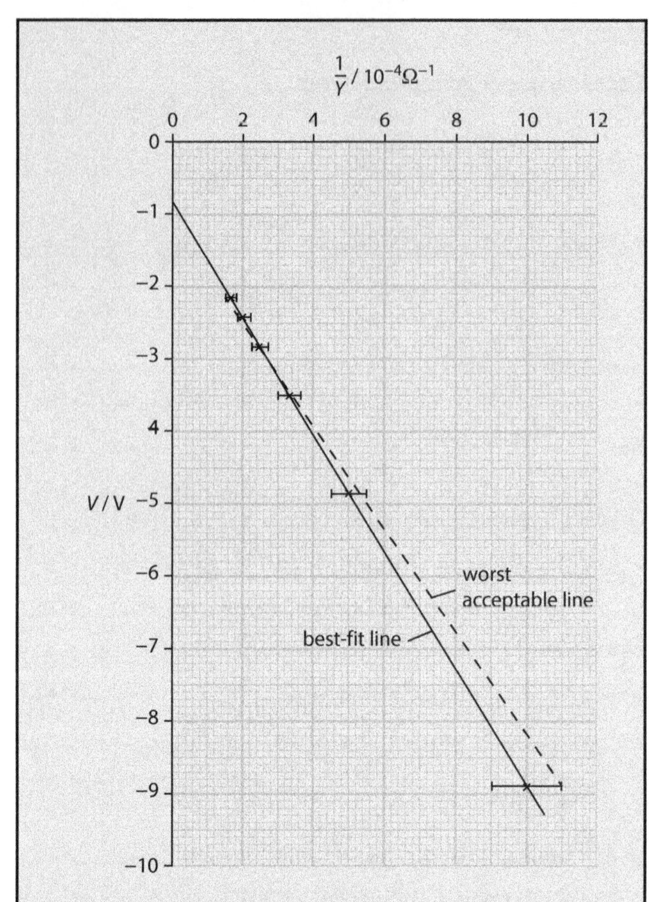

Figure 13.3

c Gradient = $-RE$

y-intercept = $\frac{RE}{X}$

e Gradient = −8100

Gradient of worst-fit line = −7000 or −9300
Uncertainty in gradient = ± 1150

y-intercept = −0.8

f y-intercept of worst-fit line = −0.43 *or* −1.1
Uncertainty in y-intercept = ± 0.33

g E = 3.0 V X = 10 kΩ

h Percentage uncertainty in E = 13.5% + 5% = 18.5%
Percentage uncertainty in X = 13.5% + 37.5% = 51% *or* 18.5% + 5% + 37.5% = 61%

i Absolute uncertainty in E = ± 0.6 V
Absolute uncertainty in X = ± 5 kΩ *or* ± 6 kΩ

j R should be measured. Circuit diagram with either an ammeter and voltmeter *or* a circuit diagram with an ohmmeter.

Chapter 14: Magnetic fields, electromagnetism and charged particles

Chapter outline

This chapter relates to Chapter 26: Magnetic fields and electromagnetism and Chapter 27: Charged particles, in the coursebook.

In this chapter learners will complete investigations on:

- 14.1 The variation of the force on a conductor in a magnetic field
- 14.2 Planning investigation into how the separation of two foils carrying a current varies with the current
- 14.3 Planning investigation into the magnetic field of a coil using a hall probe
- 14.4 How the strength of a magnetic field in a coil varies
- 14.5 Observing charged particles

Practical Investigation 14.1: The variation of the force on a conductor in a magnetic field

Skills focus

See the Skills grids at the front of this book for details of the skills developed and used in this investigation.

Duration

The practical will take about 60 minutes.

Preparing for the investigation

See details provided in Practical investigation 14.1 in the workbook.

Equipment

Each learner or group will need:

- copper wire
- pair of magnets and a yoke
- ammeter
- 2 retort stands and clamps
- connecting leads.

For this experiment a mass balance and high current power supply are needed.

Safety

There are no specific safety considerations for this practical.

Carrying out the investigation

- Learners may need help in calculating uncertainties in $(R - R_0)$ and drawing the worst acceptable line.

  Learners may also need help to determine the absolute uncertainties.

  Units may be problematic when determining B ensuring that they are in SI units. Learners may also need help in remembering to convert units both for the length L measured in centimetres and the mass measured in grams so that they do not make a power of ten error.

  Learners could also determine the (percentage) uncertainty in the value of B by a different method, e.g. a maximum / minimum method, and check that the answers are similar.

Common learner misconceptions

- Learners may need help to understand the direction of the force.

Sample results

$L = 5.1 \pm 0.1\,\text{cm}$ $R_0 = 62.2 \pm 0.1\,\text{g}$ $N = 5$

See Table 14.1

I / A	R / g	$(R - R_0)$ / g
2.0	66.1 ± 0.1	3.9 ± 0.2
2.5	67.0 ± 0.1	4.8 ± 0.2
3.0	68.1 ± 0.1	5.9 ± 0.2
3.5	69.0 ± 0.1	6.8 ± 0.2
4.0	69.9 ± 0.1	7.7 ± 0.2
4.5	70.9 ± 0.1	8.7 ± 0.2

Table 14.1

Answers to the workbook questions (using the sample results)

a See Table 14.1

b, d See Figure 14.1.

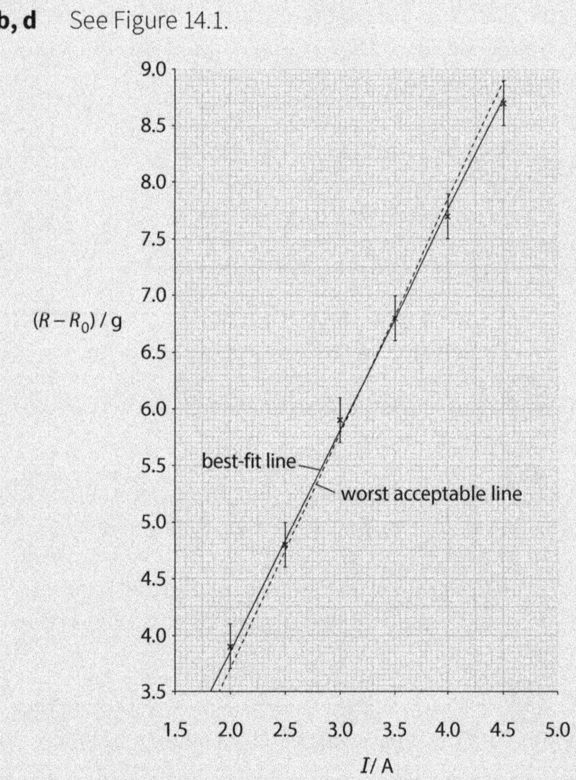

Figure 14.1

c Gradient $= \dfrac{NBL}{g}$

e Gradient = 1.95

Gradient of worst line = 1.8 or 2.12
Uncertainty in gradient = ±0.16

f $B = 0.075\,\text{T}$ *or* $\text{N A}^{-1}\,\text{m}^{-1}$ *or* Wb m^{-2}

g Percentage uncertainty in $B = \left[\dfrac{0.16}{1.95} + \dfrac{0.1}{5.1}\right] \times 100 = 10.2\%$

h The copper wire needs to be at right angles to the direction of the magnetic field. A set square or protractor could be used or the distance from the wire could be checked at each end of the magnet

i The current and/or the number of turns or wire could be increased.

Practical Investigation 14.2: Planning

Investigation into how the separation of two foils carrying a current varies with the current

Skills focus

See the Skills grids at the front of this book for details of the skills developed and used in this investigation.

Duration

The planning investigation will take about 40 minutes.

Preparing for the investigation

Learners need to:

- know about different types of variables and the control of quantities
- be able to draw graphs and calculate their gradient
- be confident in the use of logarithms.

Variables

Learners should identify:

- the dependent variable is s, the separation of the foils
- the independent variable is L, the length of the foil
- the variable to be controlled is the current in the foils. This should be kept constant. Learners may also suggest keeping the width of the foils constant.

Equipment

Each learner or group will need:

- foil
- wooden retort stand

- power supply
- ammeter
- connecting leads
- ruler
- callipers
- scissors.

Safety considerations

- There should be a precaution about the heating effect of the (high) current.

Method

A suggested method is as follows.

- Labelled diagram showing the two foils connected to a d.c. power supply. The diagram may also show a horizontal ruler fixed in a retort stand.
- The method should state that the lengths of the foils are measured.

Extra detail, for example:

- detail on measuring *s*, e.g. use callipers
- fix measuring device in a retort stand, mark the mid-points of the foils
- reason for using wooden retort stands rather than steel
- check width of foil is constant along the length of both foils
- use a high current to produce a large force
- use a conductor across the width of the foil to make good electrical contact
- repeat measurements of length at different positions and find average length
- adjust a rheostat with an ammeter in circuit to ensure that the current remains constant.

Carrying out the investigation

- Learners may need help determining the logarithmic relationship.

  Some learners will need help choosing an appropriate graph to plot.

  Learners who have finished the investigation could trial their plan and analyse their results.

Common learner misconceptions

- Learners may have difficulty realising that the foils need to be connected in series so that they repel.

Answers to the workbook questions

a Plot a graph of $\lg s$ against $\lg L$

Relationship is valid if the result is a straight line with gradient m

m = gradient

$k = 10^{y\text{-intercept}}$

Practical Investigation 14.3: Planning

Investigation into the magnetic field of a coil using a hall probe

Skills focus

See the Skills grids at the front of this book for details of the skills developed and used in this investigation.

Duration

The planning investigation will take about 40 minutes.

Preparing for the investigation

You could demonstrate the use of a Hall probe.

Learners need to:

- know about different types of variables and the control of quantities
- be able to draw graphs and calculate their gradient and *y*-intercept
- be confident in the use of natural logarithms.

Variables

Learners should identify:

- the dependent variable is *B*, the strength of magnetic field

- the independent variable is *d*, the distance from the centre of the coil
- the variable to be controlled is the current in the coil.

Equipment

Each learner or group will need:

- coil
- wooden retort stand
- power supply
- ammeter
- connecting leads
- ruler
- callipers.

Safety considerations

There should be a precaution about the heating effect of the (high) current.

Method

A suggested method is as follows.

- Labelled diagram showing the coil connected to a d.c. power supply. The diagram may also show a horizontal ruler fixed in a retort stand.
- The method should state how the distance from the centre of the coil is measured.

Extra detail, for example:

- how the Hall probe is calibrated
- rotate probe until maximum reading is obtained
- use a high current and a large number of turns per unit length on the coil to produce a large magnetic field
- detail of how to find the centre of the coil to measure *d*, e.g. measure from the end of the coil to the probe, and add on half the length of the coil. This could be further expanded by measuring the length of the coil from different positions and finding the average.
- adjust a rheostat with an ammeter in the circuit to ensure that the current remains constant.

Carrying out the investigation

- Learners may need help determining the logarithmic relationship.

Some learners will need help choosing an appropriate graph to plot.

Learners who have finished the investigation, could trial their plan and analyse their results. The value of B_0 could be compared with a theoretical calculation.

Common learner misconceptions

- Learners may have difficulties in realising that the foils need to be connected in series so that they repel.

Answers to the workbook questions

a Plot a graph of $\ln B$ against d

relationship is valid if the result is a straight line

$k = -\text{gradient}$

$B_0 = e^{y\text{-intercept}}$

Practical Investigation 14.4: Investigation into how the strength of a magnetic field in a coil varies

Skills focus

See the Skills grids at the front of this book for details of the skills developed and used in this investigation.

Duration

The practical will take about 60 minutes.

Preparing for the investigation

See details provided in Practical investigation 14.4 in the workbook.

Equipment

Each learner or group will need:

- long metal coil
- variable resistor
- ammeter
- power supply
- Hall probe
- voltmeter

- metre rule
- connecting leads.

Safety considerations

- The metallic coils may become hot.

Carrying out the investigation

- Learners may need assistance in setting up the Hall probe. Good connection will be needed to the long coil, perhaps scraping away any insulation.

Learners may also need help in the mathematical analysis. Units may be problematic when determining k.

Some learners will need to be reminded of the need to include a quantity and correct unit for the column headings in all their tables.

Learners who have finished the investigation could calibrate the Hall probe and then determine the theoretical value of k and compare it with their results.

Common learner misconceptions

- Learners may need assistance in realising that there are **two** circuits.
- Some learners will need to be reminded of the need to include a quantity and correct unit for the column headings.

Sample results

a $N = 10$ $I = 2.5 \pm 0.1$ A

See Table 14.2.

L / cm	$\frac{1}{L}$ / m^{-1}	V / mV
10.8 ± 0.4	9.26 ± 0.34	45
14.4 ± 0.4	6.94 ± 0.19	33
17.8 ± 0.4	5.62 ± 0.13	26
22.2 ± 0.8	4.50 ± 0.16	20
25.0 ± 0.8	4.00 ± 0.13	18
29.8 ± 0.8	3.36 ± 0.09	15

Table 14.2

Answers to the workbook questions (using the sample results)

a See Table 14.2

b, d See Figure 14.2.

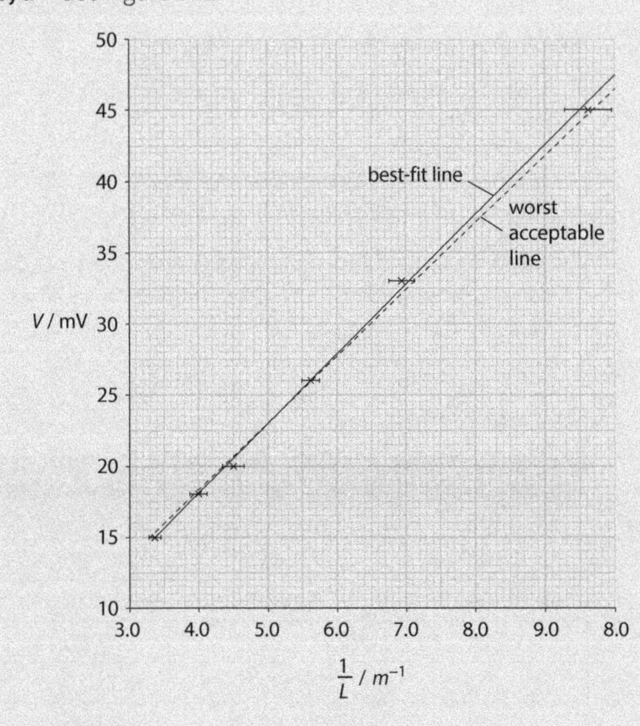

Figure 14.2

c Gradient = kNI

e Gradient = 4.8

Gradient of worst-fit line = 4.5 or 5.1
Uncertainty in gradient = ±0.3

f $k = 1.9 \times 10^{-4}$ V A^{-1}

g Percentage uncertainty in $k = \left[\frac{0.3}{4.8} + \frac{0.1}{2.5}\right] \times 100 = 10\%$

h Add marks to the coils so that L always the same. Use a set square to measure between the two marks.

i Adjust Hall probe until it reads a maximum. Repeat measurements of V with the Hall probe rotated through 180° and determine the mean value of V. Calibrate the Hall probe in a known magnetic field.

Practical Investigation 14.5: Observing charged particles investigation

Skills focus

See the Skills grids at the front of this book for details of the skills developed and used in this investigation.

Duration

The practical will take about 60 minutes.

Preparing for the investigation

This investigation is a teacher demonstration. You should set up the electron tube following the manufacturer's instructions. You should indicate the connections to the Helmholtz coils.

Equipment

Each learner or group will need:

- ruler.

Access to:

- fine beam electron tube (preferably with a vertical electron gun)
- high voltage power supply with voltmeter
- power supply for the cathode filament
- power supply for the Helmholtz coils
- shrouded connecting leads.

Safety considerations

- High voltages are being used which can cause fatal shock, so there should **not** be any bare wire connections. All leads and connectors should be rated at the voltage to be used. Teachers should have a good knowledge of high voltage electricity and the dangers of high voltage electricity.
- Learners should observe well away from the apparatus when it is being used. The electron tube is fragile (and expensive!) and should be handled carefully. It will implode if broken. Use the stand specifically designed for holding them.

Carrying out the investigation

- The teacher should set up the experiment as shown in Figure 14.3. An explanation of the cathode heating circuit, the accelerating circuit and the circuit for the magnetic coils should be given. The method to measure *d* should be discussed.

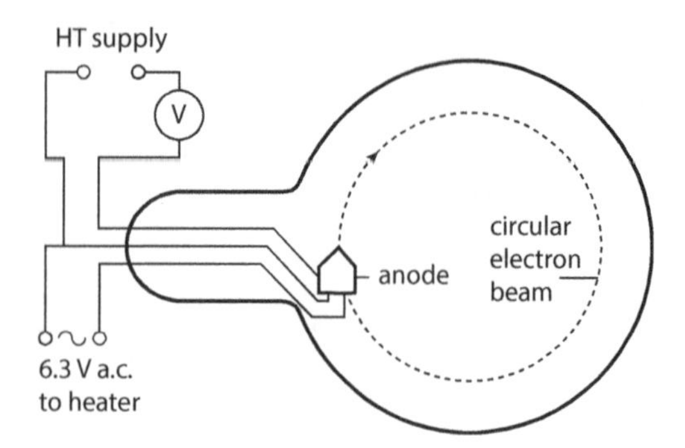

Figure 14.3

Learners may need help in the mathematical analysis.

Learners will also need to access material with the data for the rest mass of an electron and the charge on an electron.

Units may be problematic when determining *B*.

Some learners will need to be reminded of the need to include a quantity and correct unit for the column headings in all their tables.

Learners who have finished the investigation could compare their answer with the predicted value for Helmholtz coils. Alternatively, their answer could be compared with the measurement of *B* using a calibrated Hall probe.

Common learner misconceptions

- Learners may need assistance in understanding the circuits.
- Learners may need to review dealing with uncertainties with terms containing powers and square roots.
- Some learners will need to be reminded of the need to include a quantity and correct unit for the column headings.

Sample results

V / V	d / cm	d^2 / cm^2
100	1.6 ± 0.1	2.56 ± 0.32
140	1.9 ± 0.1	3.61 ± 0.38
180	2.2 ± 0.1	4.84 ± 0.44
220	2.4 ± 0.1	5.76 ± 0.48
260	2.6 ± 0.1	6.76 ± 0.52
300	2.8 ± 0.1	7.84 ± 0.56

Table 14.3

Answers to the workbook questions (using the sample results)

a See Table 14.3

b, d See Figure 14.4.

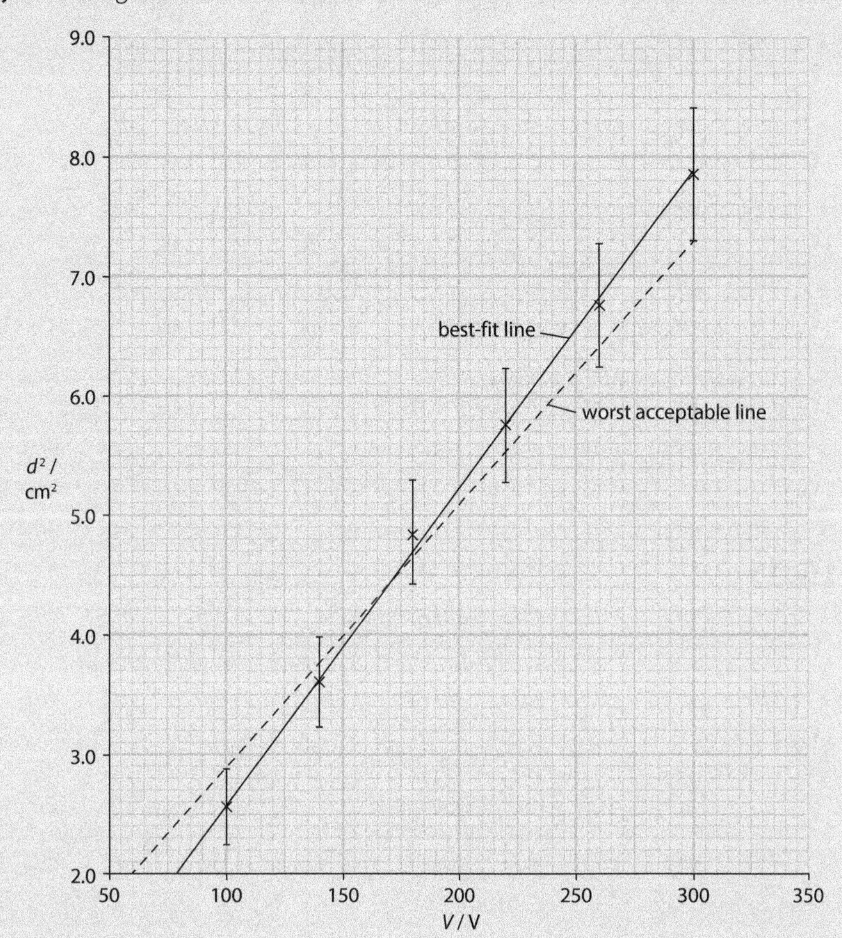

Figure 14.4

c Gradient $= \frac{8m}{B^2e}$

e Gradient = 0.026

Gradient of worst-fit line = 0.022 *or* 0.031
Uncertainty in gradient = ±0.005

f $B = 4.2 \times 10^{-3}$ T

g Percentage uncertainty in $B = \frac{1}{2} \times \left[\frac{0.005}{0.026}\right] \times 100 = 9.6\%$

h Use a bright light and measure the shadow on a screen behind the apparatus. Use two clear rulers, one on each side of the glass tube.

Chapter 15: Electromagnetic induction and alternating currents

Chapter outline

This chapter relates to Chapter 28: Electromagnetic induction and Chapter 29: Alternating currents, in the coursebook.

In this chapter learners will complete investigations on:

- 15.1 Planning investigation into the height of a metal ring above a current carrying coil
- 15.2 A bar magnet moving through a coil
- 15.3 Planning investigation into eddy currents
- 15.4 Planning investigation into the effect of the iron core of a transformer
- 15.5 Ripple voltages in a rectification circuit

Practical Investigation 15.1: Planning

Investigation into the height of a metal ring above a current carrying coil

Skills focus

See the Skills grids at the front of this book for details of the skills developed and used in this investigation.

Duration

The planning investigation will take about 40 minutes.

Preparing for the investigation

Learners need to understand that a changing magnetic field is needed so an alternating current is required for electromagnetic induction.

Learners need to:

- know about different types of variables and the control of quantities
- be able to draw graphs and calculate their gradient
- be confident in the use of logarithms.

Variables

Learners should identify:

- the dependent variable is h, the height of the aluminium ring
- the independent variable is I, the current in the coil
- the variable to be controlled is the number of turns on the coil, which should be kept constant. Learners may also suggest keeping the frequency of the a.c. supply constant.

Equipment

Each learner or group will need:

- coil
- steel retort stand
- alternating power supply or signal generator
- ammeter
- rheostat or variable power supply (to vary current)
- connecting leads
- ruler.

Safety considerations

- There should be a precaution about the heating effect of the (high) current.

Method

A suggested method is as follows.

- Labelled diagram showing the coil connected to an a.c. power supply or signal generator.
- The diagram may also show a vertical ruler fixed in a retort stand.
- The method should state that the maximum height is measured for different currents.

Extra detail, for example:

- reason for using steel/iron retort stand
- use a high current to produce a large change in magnetic flux density
- use a high frequency current to increase the rate of change of flux density
- repeat measurements of height at different positions and find average height
- adjust a rheostat with an ammeter in circuit to ensure that the current remains constant.

Carrying out the investigation

- Learners may need help determining the logarithmic relationship.

  Some learners will need help choosing an appropriate graph to plot.

  Learners who have finished the investigation could trial their plan and analyse their results.

Common learner misconceptions

- Learners may have difficulties in realising that an alternating power supply is needed.

Answers to the workbook questions

a Plot a graph of $\lg h$ against $\lg I$

relationship is valid if the result is a straight line with gradient q

q = gradient

$p = 10^{y\text{-intercept}}$

Practical Investigation 15.2: A bar magnet moving through a coil

Skills focus

See the Skills grids at the front of this book for details of the skills developed and used in this investigation.

Duration

The practical will take about 60 minutes.

Preparing for the investigation

See details provided in Practical investigation 15.2

Equipment

Each learner or group will need:

- bar magnet
- coil (50–100 turns)
- trolley
- inclined plane
- light gate
- voltmeter or cathode ray oscilloscope
- timer
- ruler
- connecting leads
- retort stands.

Safety

There are no specific safety issues for this investigation.

Carrying out the investigation

- Learners may need help in calculating uncertainties in average t and $\frac{1}{t}$ and the drawing of the worst acceptable line.
- Learners may also need assistance with understanding the light gate method for determining the time.

Learners may also need help to determine the absolute uncertainties.

Some learners will need to be reminded of the need to include a quantity and correct unit for the column headings in all their tables.

Learners could derive the given relationship. Learners could also consider the physical significance of the y-intercept.

Common learner misconceptions

- Learners may need reminding that t/ms is the time measured in milliseconds.
- Learners may find the unit for k difficult.

Sample results

$L = 5.2 \pm 0.1$ cm

See Table 15.1.

t_1 / ms	t_2 / ms	average t / ms	$\frac{1}{t}$ / s^{-1}	E / mV
90	82	86±4	11.6±0.5	11
62	54	58±4	17.2±1.2	16
50	42	46±4	21.7±1.9	21
39	31	35±4	28.6±3.3	26
33	25	29±4	34.5±4.8	32
20	28	24±4	41.7±7.1	40

Table 15.1

Answers to the workbook questions (using the sample results)

a, b See Table 15.1.

c, e See Figure 15.1.

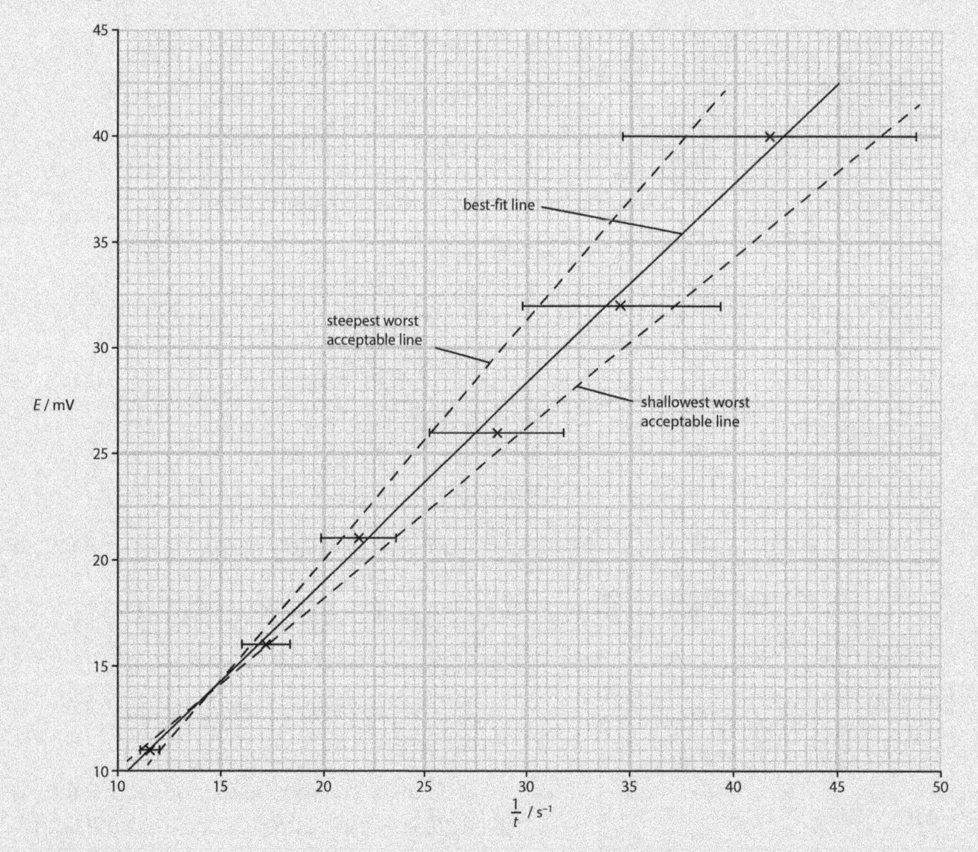

Figure 15.1

Note for the sample data, the line of best fit should **not** pass through both the top and bottom plots.

d Gradient = kL^2

f Gradient = 0.95×10^{-3}

Gradient of worst-fit line = 0.74 *or* 1.12
Uncertainty in gradient = $\pm 0.19 \times 10^{-3}$

g $k = 0.35\,\text{V s m}^{-2}$

h Percentage uncertainty in
$k = \left(\frac{0.19}{0.95} + 2 \times \frac{0.1}{5.2}\right) \times 100 = 24\%$

i Absolute uncertainty in $k = 0.32 \times 0.31 = 0.099$

j t is a very small value. L could be made larger so that t will be larger, thus reducing the percentage uncertainty in t. In effect, an instantaneous average speed is being measured since the trolley is accelerating.

k It is difficult to determine the maximum value from a voltmeter. The output could be displayed on a storage oscilloscope or data logger. The resulting graph would then enable the maximum value of E to be determined.

Practical Investigation 15.3: Planning

Investigation into eddy currents

Skills focus

See the Skills grids at the front of this book for details of the skills developed and used in this investigation.

Duration

The planning investigation will take about 40 minutes.

Preparing for the investigation

Learners need to:

- understand eddy currents
- know about different types of variables and the control of quantities
- be able to draw graphs and calculate their gradient and y-intercept
- be confident in the use of natural logarithms.

Variables

Learners should identify:

- the dependent variable is t, the time for the copper sheet to stop oscillating
- the independent variable is w, the distance from the centre of the coil
- the variables to be controlled are keeping the depth of the gap constant and ensuring that both values of w are the same for each different sized gap. The sheet should also be released from the same position.

Equipment

Each learner or group will need:

- metal cutters/pliers
- wooden retort stand
- stopwatch
- horseshoe magnet
- ruler or callipers.

Safety considerations

- There should be a precaution regarding the use of metal cutters and sharp edges.

Method

A suggested method is as follows.

- Labelled diagram showing the sheet pivoted. The diagram should also show the pivot supported in a retort stand.
- The method should state releasing the sheet from rest at the same point each time. Stopwatch started when the sheet is released and stopped when sheet comes to rest.

Extra detail, for example:

- repeat experiment for each value of *d* and find the average *t*
- use a strong magnet
- method to ensure that the sheet is aligned with the coils and passes at right angles to the direction of the magnetic field
- avoid draughts
- use Vernier callipers to measure *d*
- detail on starting the sheet from the same position each time
- measure *d* at several positions and find the average.

Carrying out the investigation

- Learners may need help determining the logarithmic relationship.

  Some learners will need help choosing an appropriate graph to plot.

  Learners who have finished the investigation could trial their plan and analyse their results.

Common learner misconceptions

- Learners may have difficulties in understanding the position of the magnet in relation to the sheet.

> **Answers to the workbook questions**
>
> **a** Plot a graph of ln *t* against *d*
>
> relationship is valid if the result is a straight line
>
> $q = -$ gradient
>
> $k = e^{y\text{-intercept}}$

Practical Investigation 15.4: Planning

Investigation into the effect of the iron core of a transformer

Skills focus

See the Skills grids at the front of this book for details of the skills developed and used in this investigation.

Duration

The planning investigation will take about 40 minutes.

Preparing for the investigation

Learners need to:

- understand transformers
- know about different types of variables and the control of quantities
- be able to draw graphs and calculate their gradient and *y*-intercept
- be confident in the use of natural logarithms.

Variables

Learners should identify:

- The dependent variable is *E*, the e.m.f. across the secondary coil.
- The independent variable is *d*, the distance that the iron bar does not overlap the C-core.
- The variables to be controlled are: the current flowing in the primary coil, the number of turns on each coil and the frequency of the a.c. supply. These must be kept constant.

Equipment

Each learner or group will need:

- two coils
- iron bar
- a.c. power supply or signal generator
- voltmeter or cathode ray oscilloscope
- ammeter
- rheostat
- connecting leads
- ruler or callipers.

Safety considerations

- There should be a safety precaution about the heating effect of the iron core.

Method

A suggested method is as follows.

- Labelled diagram showing the primary coil connected to an a.c. power supply or signal generator and the secondary coil connected to the voltmeter or cathode ray oscilloscope.
- The method should state how the distance *d* is measured.

Extra detail, for example:

- use a high current, a large number of turns and a large magnetic field
- detail on measuring *d*, e.g. measure from both sides of the C-core and determine an average value
- adjust a rheostat with an ammeter in the circuit to ensure that the current remains constant
- how to determine the e.m.f. across the secondary from a cathode ray oscilloscope.

Carrying out the investigation

- Learners may need help determining the logarithmic relationship.

  Some learners will need help choosing an appropriate graph to plot.

  Learners who have finished the investigation could trial their plan and analyse their results.

Common learner misconceptions

- Learners may have difficulties in realising that there are two separate circuits with a transformer and that the primary coil should be connected to an a.c. power supply.

Answers to the workbook questions

a Plot a graph of ln *E* against *d*

relationship is valid if the result is a straight line

$k = -\text{gradient}$

$E_0 = e^{y\text{-intercept}}$

Practical Investigation 15.5: Investigation into ripple voltages in a rectification circuit

Skills focus

See the Skills grids at the front of this book for details of the skills developed and used in this investigation.

Duration

This practical will take about 60 minutes.

Preparing for the investigation

See details provided in the workbook for this investigation.

Learners need to be confident:

- in measuring potential difference from a cathode ray oscilloscope
- in determining frequency from a cathode ray oscilloscope.

Equipment

Each learner or group will need:

- capacitor
- diode
- various resistors in the range 3.3 kΩ to 10 kΩ
- a.c. power supply
- connecting leads
- cathode ray oscilloscope.

Safety

There are no specific safety issues for this investigation.

Carrying out the investigation

- Check the circuit if it is set up by learners.
- If an electrolytic capacitor is used then the polarity should be checked.

  Learners may need help in measuring V_{min} and determining the frequency from the cathode ray oscilloscope.

  Learners may also need help in the mathematical analysis. In particular, determining the gradient with the 10^{-3} power term on the *x*-axis.

  Some learners will need to be reminded of the need to include a quantity and correct unit for the column headings in all their tables.

Learners who have finished the investigation could repeat the experiment using a full bridge rectifier.

Learners could also determine the absolute uncertainty in C using a maximum/minimum method.

- Learners may need help in determining voltages from the oscilloscope.
- Learners may need help in determining the frequency of the power supply from the oscilloscope using the time-base and $f = \frac{1}{T}$.

Common learner misconceptions

- Learners may need assistance in setting up the circuit.
- Learners may need to review dealing with uncertainties.

Sample results

Frequency, $f = 50\,\text{Hz}$ $V_{max} = 9.2 \pm 0.1\,\text{V}$

See Table 15.2.

R / kΩ	$\frac{1}{R}$ / $10^{-3}\,\Omega^{-1}$	V_{min} / V	V_R / V
3.3	0.303	3.5 ± 0.1	5.7 ± 0.2
4.7	0.213	5.3 ± 0.1	3.8 ± 0.2
5.5	0.182	5.9 ± 0.1	3.3 ± 0.2
6.8	0.147	6.5 ± 0.1	2.7 ± 0.2
7.6	0.132	6.8 ± 0.1	2.4 ± 0.2
9.6	0.104	7.4 ± 0.1	1.8 ± 0.2

Table 15.2

Answers to the workbook questions (using the sample results)

a, b See Table 15.2

c, e See Figure 15.2.

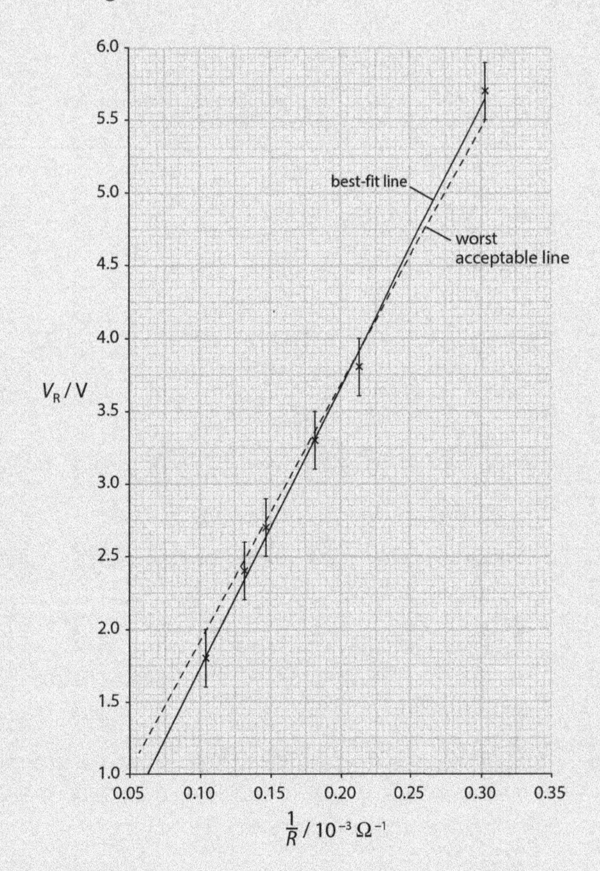

Figure 15.2

d Gradient $= \frac{V_{max}}{fC}$

f Gradient = 19 200

Gradient of worst-line = 17 600 *or* 21 600
Uncertainty in gradient = ±2000

g $C = 9.6 \times 10^{-6}$ F

h Percentage uncertainty in
$C = \left(\frac{2000}{19200} + \frac{0.1}{9.2}\right) \times 100 = 11.5\%$

i Absolute uncertainty in $C = 0.115 \times 9.6 \times 10^{-6} =$
1.1×10^{-6} F

Maximum/minimum methods:

$\max C = \frac{9.3}{50 \times 17\,600} = 10.6\ \mu\text{F}$

$\min C = \frac{9.1}{50 \times 21\,600} = 8.4\ \mu\text{F}$

j *R* could be determined either by a voltmeter and ammeter method or by connecting the resistance combination to an ohm-meter.

k The trace on the oscilloscope screen should be as large as possible.
V_{max} could be determined and then the trace could be adjusted so that that V_{max} is positioned at the top of the screen and V_{min} at the bottom of the screen. This means would reduce the percentage uncertainty in V_R.

Chapter 16:
Quantum physics, nuclear physics and medical imaging

Chapter outline

This chapter relates to Chapter 30: Quantum physics and Chapter 31: Nuclear physics and medical imaging, in the coursebook.

In this chapter learners will complete investigations on:

- 16.1 Determining Planck's constant
- 16.2 Data analysis investigation into measuring radioactive decay constant
- 16.3 Planning investigation into X-ray attenuation
- 16.4 Data analysis investigation into the Larmor frequency

Practical Investigation 16.1: Determining Planck's constant

Skills focus

See the Skills grids at the front of this book for details of the skills developed and used in this investigation.

Duration

The practical work will take 30 minutes; the analysis and evaluation questions will take 30 minutes.

Preparing for the investigation

Learners should be able to:

- should know the equation for the energy of a photon and the relationship between frequency and wavelength
- set up a potentiometer circuit
- combine percentage uncertainty
- measure the smallest voltage that just causes LEDs of different colour to emit light.

Equipment

Each learner or group will need:

- low voltage d.c. power supply of at least 3 V: two 1.5 V cells in series is sufficient
- LEDs of several different colours, ideally from violet to red
- safety resistor of a few hundred ohms
- variable resistor to be connected as a potentiometer; the value of this resistor is not critical but it must allow the voltage across the LED to be altered from 0 to a few volts
- digital multimeter / voltmeter reading 2 V to at least 0.01 V
- small opaque tube, e.g. black card to be placed over the LED.

Access to:

- the internet or a colour chart showing the wavelength of light of different colours.

Alternative equipment

- If the power supply has a variable voltage that can be adjusted continuously the variable resistor may not be needed or may be added in series with the safety resistor.

Safety considerations

- The safety resistor of a few hundred ohms is included to ensure that the LED cannot be provided with too much current which might make it fail. You may need to adjust the value of this resistance in accord with the maximum voltage of the power supply which should not need to exceed a few volts.

Carrying out the investigation

- Learners need to find the smallest voltage across the LED which just enables it to glow. This is achieved by placing one end of the tube over the LED and looking down the tube from the other end. It is helpful to find this voltage just as light can be seen and just as light cannot be seen when the voltage is increased and when it is decreased, respectively. There is an element of judgment in this and learners should not expect to always achieve the same value. The different values obtained allow for a realistic calculation of the uncertainty.

Teachers may need to give learners confidence in finding the minimum voltage by demonstrating one of the readings.

Learners can use the values of uncertainty in their values of wavelength to plot error bars on the *x*-axis as well as on the *y*-axis. They can also suggest reasons for systematic error and the effect of any systematic error that they suggest on the result of the experiment.

Common learner misconceptions

- Learners may find difficulty with powers of ten for wavelength and its inverse, and a factor of 10^{-7} has been left within the table rather than being incorporated within the heading of the columns.

Sample results

Table 16.1 gives an idea of the results the learners should end the investigation with.

LED	Voltmeter reading / V				λ / m	$\frac{1}{\lambda}$ / m^{-1}
	1st	2nd	3rd	average		
1	0.72	0.71	0.73	0.72 ± 0.01	5.78×10^{-7}	1.73×10^{6}
2	0.85	0.84	0.86	0.85 ± 0.01	5.46×10^{-7}	1.83×10^{6}
3	1.44	1.42	1.46	1.44 ± 0.02	4.36×10^{-7}	2.29×10^{6}
4	1.67	1.64	1.70	1.67 ± 0.03	4.05×10^{-7}	2.47×10^{6}
5	1.98	1.99	2.03	2.00 ± 0.03	3.66×10^{-7}	2.73×10^{6}

Table 16.1

Answers to the workbook questions (using the sample results)

a See Table 16.1

b See Figure 16.1

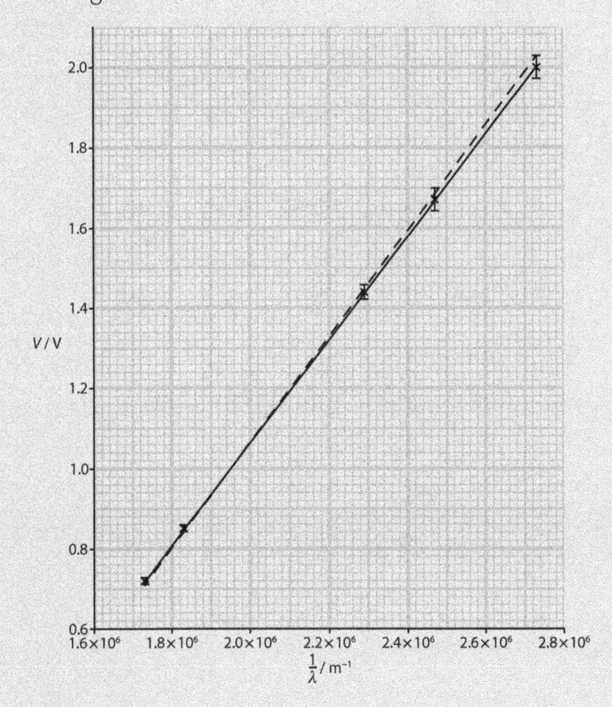

Figure 16.1

c Gradient of best-fit line = 1.28×10^{-6}

Gradient of worst-fit line = 1.31×10^{-6} (V m) approx

d Gradient = $\frac{hc}{e}$

e h = gradient × $\frac{e}{c}$ = 6.8×10^{-34} J s

f Worst value of $h = 7.0 \times 10^{-34}$ J s so uncertainty = $\pm 0.2 \times 10^{-34}$ J s

g The value obtained is, within the limits of uncertainty, consistent with the accepted value 6.6×10^{-34} J s.

h Possible causes of systematic error may be that the eye has different responses at different wavelengths and may, for example, require brighter light at blue wavelengths to detect the light, meaning that *V* is larger at blue wavelengths and causing a steeper graph. Uncertainty in a value of *V* may be due to the difficulty of making small changes in voltage and this will depend on the quality of your potential divider.

Practical Investigation 16.2: Data analysis

Investigation into measuring a radioactive decay constant

Skills focus

See the Skills grids at the front of this book for details of the skills developed and used in this investigation.

Duration

The data analysis and evaluation questions will take about 40 minutes.

Preparing for the investigation

It may be helpful for teachers to show a simulation of radioactive decay, e.g. https://phet.colorado.edu/en/simulation/legacy/radioactive-dating-game or use dice or coins to illustrate radioactive decay and to draw similar ln graphs to calculate the decay constant and then calculate half-life.

Learners need to:

- be able to handle natural logarithms and know the relationship $\ln e^a = a$
- draw ln graphs and insert error bars.

Learners will calculate a count rate corrected for background and plot a logarithmic graph to calculate the radioactive decay constant.

Equipment

There is no practical equipment necessary in this data analysis exercise. Some manufacturers provide a 'prepared protactinium generator' apparatus for this experiment, with the uranium salt and alcohol in a sealed plastic bottle. Details are available at http://practicalphysics.org/measuring-half-life-protactinium.html

If apparatus is available for demonstration, strict safety precautions must be taken to avoid learners handling sources or being close.

Safety considerations

Learners may suggest that, in performing the experiment, the source must be far from the experimenter, that the bottle should be stored safely in a lead container away from any contact with people and should only be used for as short a time as possible. In this experiment the source is in a sealed container and a tray should collect any spillage. The experimenter may use gloves in case of any spillage and it might be sensible to shield the lower part of the bottle with lead.

Carrying out the investigation

- Learners need to convert the background count of 50 in 100 s to 5 in 10 s before subtracting this value from the counts in the table. They then need to convert this count into a rate per second. This two-stage process may cause difficulties and may have to be explained to some learners, but hopefully groups of learners, once faced with the problem, will work out their own solution.

  Learners who are not studying mathematics may need help with the logarithm of an exponential quantity.

  Learners can also be asked to find the half-life $t_{\frac{1}{2}}$ of the source by using the equation $t_{\frac{1}{2}} = \frac{0.693}{\lambda}$ and also by plotting a graph of count rate against time. The uncertainty can be found in both methods and the values compared.

Common learner misconceptions

- Learners may have difficulty in working out the corrected background count rate.

Sample results

The data in the workbook provides sample results.

Answers to the workbook questions

a, b See Table 16.2.

Time t / s	Count in 10 s	Corrected count rate C / s^{-1}	ln (C / s^{-1})
0	123 ± 11	11.8 ± 1.1	2.47 ± 0.09
20	100 ± 10	9.5 ± 1.0	2.25 ± 0.10
40	80 ± 9	7.5 ± 0.9	2.01 ± 0.11
60	68 ± 8	6.3 ± 0.8	1.84 ± 0.12
80	52 ± 7	4.7 ± 0.7	1.55 ± 0.14
100	45 ± 7	4.0 ± 0.7	1.39 ± 0.16
120	34 ± 6	2.9 ± 0.6	1.06 ± 0.19
140	31 ± 6	2.6 ± 0.6	0.96 ± 0.21

Table 16.2

c See Figure 16.2

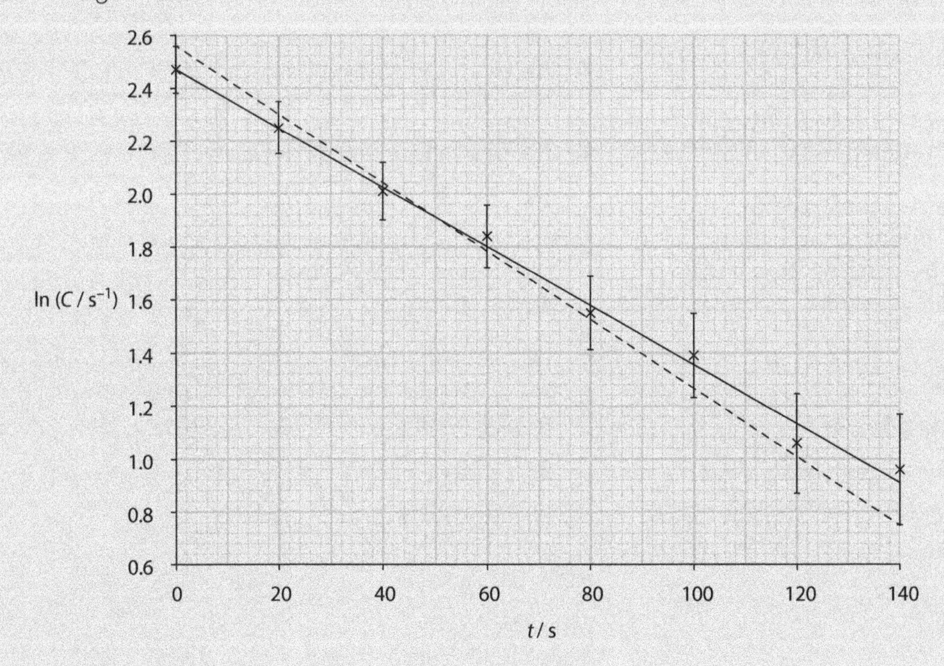

Figure 16.2

d Gradient of line of best fit = −0.011, gradient of line of worst fit = −0.013 (s^{-1})

e $\ln C = \ln C_0 - \lambda t$

f $\lambda = 0.011\ (\pm 0.02)\ \mathrm{s}^{-1}$

g The oil dissolves the protactinium present at one instant and the experiment then records the decay of this protactinium. Without the use of the oil, protactinium is continuously being created by the decay of uranium and this decay is also being recorded as well as the decay of a number of other products in the series of decay from the initial uranium.

Practical Investigation 16.3: Planning

Investigation into X-ray attenuation

Skills focus

See the Skills grids at the front of this book for details of the skills developed and used in this investigation.

Duration

The planning will take about 40 minutes.

Preparing for the investigation

Learners will need to:

- know about the attenuation coefficient for X-rays

- be able to use logarithms to base e and to use the relationships:

$$\ln(e^a) = a \text{ and } \ln(a^b) = b \ln a$$

Learners will plan an experiment to show how the count rate measured on one side of a piece of lead depends on the thickness of the lead and measure the attenuation coefficient.

Carrying out the investigation

- Learners might need help in calculating a count rate and correcting for background count. They also may not realise that there is a background count due to background radiation even when the X-ray machine is turned off.

  Learners may have difficulty in understanding that the axis of a logarithmic graph is given a label such as ln (C / s^{-1}) with the unit s inside the bracket but the unit of the gradient of a graph of ln (C / s^{-1}) against t / m is just m^{-1}.

  If more able learners have finished the investigation, they can also use an analogy with radioactive decay and suggest how the thickness of lead to reduce the count to one half can be found from the attenuation coefficient.

  You can also show or give students readings of corrected count rate and lead thickness taken from a video, obtained, for example, by searching for "gamma absorption and lead" in the video section of a browser or on YouTube.

Common learner misconceptions

- Learners might think that a medical physicist wears a lead suit. In practice, only a small lead overjacket is worn.

Variables

Learners should identify:

- dependent variable: count rate
- independent variable: thickness of sheet t.
- variables to be controlled:
  - distance from source to GM tube
  - activity of X-ray source
  - angle between axis of GM tube and X-ray source
  - time for which each reading of the count is taken.

Equipment

Each learner or group will need:

- stopwatch
- micrometer screw gauge or callipers.

Safety precautions

- When the X-ray source is switched on the medical physicist must be as far away as possible and, quite likely, behind a lead wall. The source can also be shielded with lead to avoid affecting any other person nearby.

Method

A suggested method is:

1. Measure the thickness of the lead sheet t with a micrometer or callipers.
2. Insert the sheet between the GM tube and source which are fixed in position.
3. Switch on the X-ray source and measure the count in as long a time t as reasonable, using the stopwatch.
4. Switch off and repeat the measurements for the same time and using different thicknesses of lead sheet.
5. Use 10 different thicknesses from 1 mm to 3 cm, e.g. 1 mm, 2 mm, 3 mm, 4 mm, 5 mm, 8 mm, 10 mm, 15 mm, 20 mm, 30 mm.
6. Switch the source off and measure the background count on the GM tube and counter for the same time that the measurements were made.
7. The count rate C = (the count − background count) /time for each reading.

Results

Time of count T = __________
Background count in this time = __________

See Table 16.3.

Thickness t / mm	Count in time T / s	Count C corrected for background radiation / s^{-1}	ln (C / s^{-1})

Table 16.3

Extra detail

- Since 5 mm halves the count rate a sensible maximum thickness would be, for example, 3 cm when the count rate is $\left(\frac{1}{2}\right)^6$, a reduction of 64 times. Other sensible maximum thicknesses can be suggested.
- The GM tube and source are fixed in position about 5 cm apart. This is as close as possible given that the maximum thickness of lead sheet is 3 cm. Having them close will mean that the air absorbs less radiation.
- Counting for a long time reduces the percentage uncertainty in the count and increases accuracy in the final value of the decay constant.

Answers to the workbook questions

a $\ln C = \ln C_0 - \alpha t$

b *x*-axis: thickness *t* *y*-axis: $\ln C$

c See Figure 16.3.

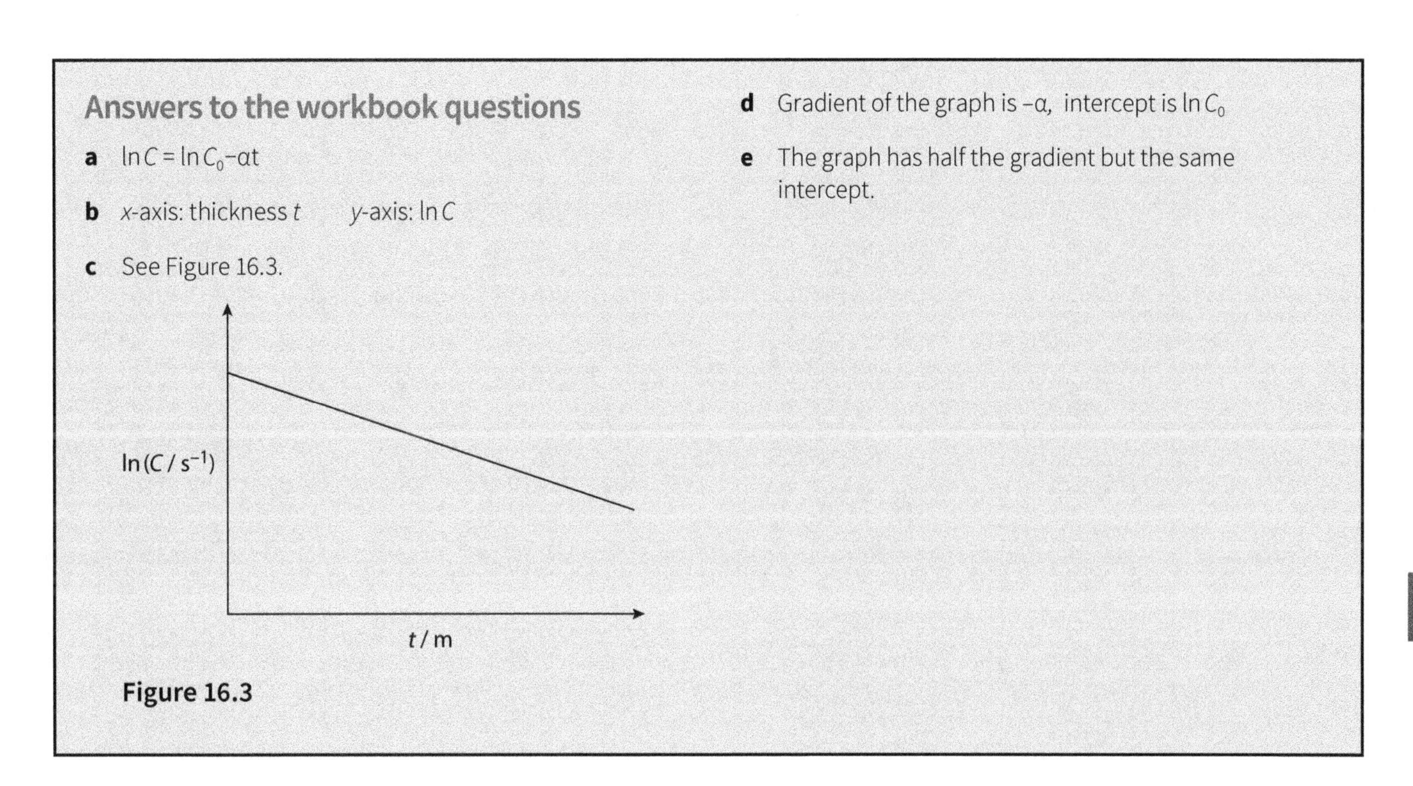

Figure 16.3

d Gradient of the graph is $-\alpha$, intercept is $\ln C_0$

e The graph has half the gradient but the same intercept.

Practical Investigation 16.4: Data analysis

Investigation into the Larmor frequency

Skills focus

See the Skills grids at the front of this book for details of the skills developed and used in this investigation.

Duration

The analysis and evaluation questions will take about 45 minutes.

Preparing for the investigation

- It is helpful if learners have studied NMR, as some of the later questions refer to the process of medical scanning by NMR. The simulation of the process can be shown in a simulation, e.g. at https://phet.colorado.edu/en/simulation/mri; here, learners can not only see the excitation and procession but can themselves see the experiment and determine the Larmor frequency of hydrogen in different magnetic field strengths. It may be sensible for learners to use this simulation before attempting the investigation.
- Learners will use data for the Larmor frequency obtained at different magnetic field strengths to determine the gyromagnetic ratio for a nucleus containing a single proton and another nucleus containing a neutron and a proton. This enables them to make a comment that the nucleus also acts as a rotating magnet even though it is uncharged.

In part **d**, learners have to compare the magnetic moment of the neutron and the proton. The combined magnetic moment of the proton and neutron is less than that of the proton alone as the neutron has a magnetic effect of the opposite sign and acts in the opposite direction. This may need to be explained to learners as it will not be immediately obvious. Neither will be the idea that spinning objects with charge act as magnets.

More able learners who have finished the investigation can obtain the Larmor frequency of other nuclei using the simulation and use the simulation to increase their knowledge of the principles of nuclear magnetic resonance imaging.

Answers to the workbook questions

a See Figure 16.4

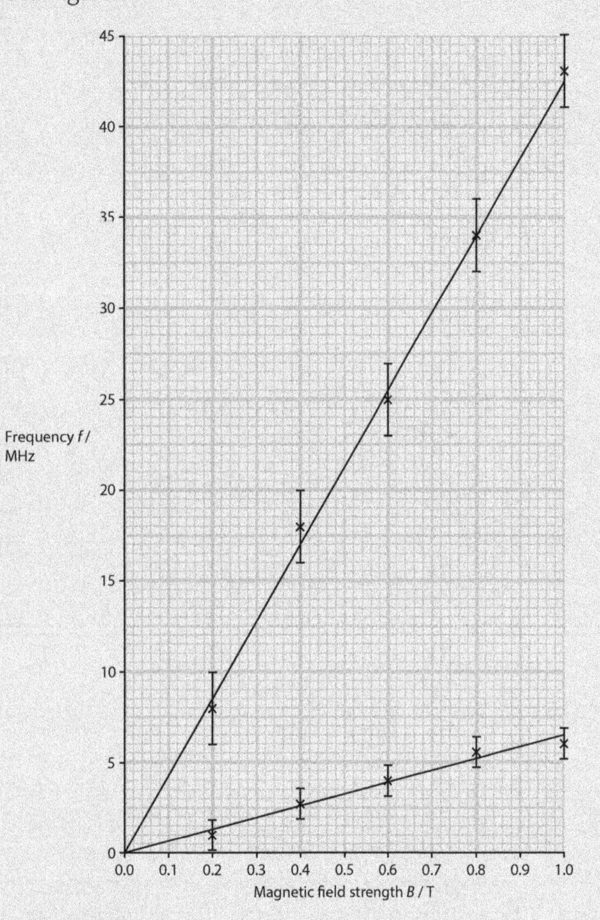

Figure 16.4

b For nuclei with one proton:

Gradient of line of best fit = 43×10^6

Gradient of line of worst fit = 45×10^6 ($Hz\,T^{-1}$)

For nuclei with one proton and one neutron:

Gradient of line of best fit = 6.5×10^6

Gradient of line of worst fit = 7.0×10^6

c Gyromagnetic ratio for ^1_1H = gradient × 2π = $2.70 \times 10^8 \pm 0.12 \times 10^8\,Hz\,T^{-1}$ *or* $s^{-1}\,T^{-1}$

Gyromagnetic ratio for $^2_1H = 41 (\pm 3) \times 10^6\,Hz\,T^{-1}$ *or* $s^{-1}\,T^{-1}$

d The strength of the neutron magnetic moment is weaker than that of the proton and in the opposite direction, as it partially, but not completely, cancels the magnetic moment of the proton.

e By measuring the Hall voltage from a Hall effect probe, which is calibrated by placing it in a known magnetic field.

f The relaxation time differs for the two environments and so the decay of the nuclei and the time taken for the signal in the receiver to decay is different in water or in tissue.